Applied Continuum Mechanics for Thermo-Fluids

Applied Continuum Mechanics for Thermo-Fluids presents the tensor notation rules and integral theorems before defining the preliminary concepts and applications of continuum mechanics. It bridges the gap between physical concepts and mathematical expressions with a rigorous mathematical treatment. After discussing fundamental concepts of continuum mechanics, the text explains basic subjects such as the Stokes hypothesis, the second coefficient of viscosity, the non-Newtonian fluids, the non-symmetric stress tensor, and the full Navier–Stokes equation.

With coverage of interdisciplinary topics, the book highlights issues such as relativistic fluid mechanics, stochastic mechanics, fractional calculus, nanoscale fluid mechanics, polar fluids, electrodynamics, and traffic flows. It describes fundamental concepts of vorticity dynamics, including the definition of vorticity and circulation, with corresponding balance equations and related theorems.

This text is intended for upper-level undergraduate and postgraduate mechanical, chemical, aerospace, civil engineering, and physics students taking continuum mechanics, advanced fluid mechanics, convective heat transfer, turbulence, or any other similar courses. In addition, this book can be an excellent resource for scientists who want to trigger research on topics related to thermo-fluids.

Instructors will be able to utilize a solutions manual and figure slides for their courses.

Applied Continuum Mechanics for Thermo-Fluids

Jafar Ghazanfarian

CRC Press
Taylor & Francis Group
Boca Raton London New York

CRC Press is an imprint of the
Taylor & Francis Group, an **informa** business

Designed cover image: Bisams, Shutterstock

First edition published 2024
by CRC Press
2385 NW Executive Center Drive, Suite 320, Boca Raton FL 33431

and by CRC Press
4 Park Square, Milton Park, Abingdon, Oxon, OX14 4RN

CRC Press is an imprint of Taylor & Francis Group, LLC

© 2024 Jafar Ghazanfarian

Library of Congress Cataloging-in-Publication Data
Names: Ghazanfarian, Jafar, author.
Title: Applied continuum mechanics for thermo-fluids / Jafar Ghazanfarian.
Description: First edition.	Boca Raton, FL : CRC Press, 2024.	Includes bibliographical references and index.
Identifiers: LCCN 2023053928 (print)	LCCN 2023053929 (ebook)	ISBN 9781032719382 (hbk)	ISBN 9781032719399 (pbk)	ISBN 9781032719405 (ebk)	ISBN 9781032724843 (eBook+)
Subjects: LCSH: Fluids--Thermal properties.	Thermofluids--Mathematical models.	Continuum mechanics.
Classification: LCC QC145.4.T5 G46 2024 (print)	LCC QC145.4.T5 (ebook)	DDC 532--dc23/eng/20240224
LC record available at https://lccn.loc.gov/2023053928
LC ebook record available at https://lccn.loc.gov/2023053929

ISBN: 978-1-032-71938-2 (hbk)
ISBN: 978-1-032-71939-9 (pbk)
ISBN: 978-1-032-71940-5 (ebk)
ISBN: 978-1-032-72484-3 (eBook+)

DOI: 10.1201/9781032719405

Typeset in Nimbus Roman
by KnowledgeWorks Global Ltd.

Access the Instructor Resources: routledge.com/cw/ghazanfarian

Dedication

To the love of my life Zahra and my beautiful always-for-me little girl Setareh.

Contents

Preface

Matter consists of atoms. In continuum mechanics, we are trying to ignore atoms. So, we have to design mathematical frameworks to present governing equations for different applications from elastic solids to superfluids. The good news is that we can build such frameworks, but the bad news is that we cannot find unified universal relations for all applications. Hence, we have to derive a different set of equations for each physical problem. On the other hand, atomistic/molecular models have their own limitations such as high computational costs and low accuracy to capture quantities such as inter-molecular forces especially in real-world applications.

We are facing a fundamental question: what is continuum mechanics? A simple answer: the derivation and investigation of a collection of governing equations based on conservation laws that have been customized regarding constitutive relations. Another strategy to find an answer to the question is to clarify what is not continuum mechanics. Thus, we need to understand non-continuum effects.

In addition, there are few textbooks on continuum mechanics that widely focus on thermo-fluid sciences and other connected fields of science or modern topics. About five years ago, I decided to start writing this book to fill the gap. The present book has been designed as a textbook for graduate students of mechanical engineering, chemical engineering, civil engineering, aerospace engineering, material sciences, condensed matter physics, and other related fields. It also can be a reference for understating vague concepts even for undergraduate students or high-level motivated researchers.

The objectives and goals are:

- Explaining tensor notation concepts in an easy-to-learn and self-learnable manner,
- Paying special attention to the organization of chapters and presentation of topics in the form of successive items or steps,
- Emphasizing applied aspects,
- Making a bridge between physical concepts and mathematical expressions,
- Collecting references from classical papers to newly published articles,
- Covering diverse applications that can be related to thermo-fluid sciences,
- Putting everything in a nutshell as much as possible,
- Helping readers find their fields of interest and trigger research,
- Introducing modern or interdisciplinary topics such as relativistic fluid mechanics, quantum fluid mechanics, machine learning tools, virtual reality, stochastic mechanics, fractional calculus, etc.

Chapter 1 presents the tensor notation rules and integral theorems. Chapter 2 tries to well define preliminary concepts and outlooks of continuum mechanics. Chapters 3 and 4 deeply investigate the derivation of conservation of mass, constitutive relations, primary and secondary forms of conservation of linear/angular momentum as well as the boundary conditions. Chapter 5 is devoted to the energy and entropy balances and their alternative forms such as the exergy equation. Chapter 6 collects a description of several diverse applications that can be connected to thermo-fluid sciences from thermo-elasticity to quantum fluids. Chapter 7 describes fundamental concepts of vorticity dynamics including the definition of vorticity and circulation and corresponding balance equations and related theorems.

I wish to express my appreciation to Dr. Zahra Shomali, Dr. Mohammad Taghilou, Dr. Robert Kovacs, and Dr. Mojtaba Ayatollahi (first two chapters) for sharing their useful comments and comprehensive discussions about different sections of the book. I kindly ask readers to send their comments on how to improve the quality of discussions, presentations, or organizations and their suggestions for adding new sections.

Jafar Ghazanfarian
University of Zanjan
2023

About the Author

Dr. Jafar Ghazanfarian has been an associate professor of mechanical engineering since 2016, assistant professor since 2011, and lecturer since 2007 at the University of Zanjan. He has taught continuum mechanics for six years to thermo-fluids graduate students. So far, he has published more than 40 full papers in high-quality WOS-indexed journals, 26 conference papers, and 2 book chapters published by CRC Press and IGI Global. He also served as referee for a wide range of journals. His publications have collected 1165 citations with an h-index of 22 and i10-index of 33 based on the GoogleScholar database. He is the chair of the Complex Heat and Flow Simulation (CHFS) research group, which currently works on developing non-Fourier models, nanoscale heat transport, modern computational fluid dynamics, moving boundary problems, machine learning tools in mechanical engineering, and renewable energies. He also has recorded over 850 hours of free educational video lectures on different topics of mechanical engineering (in Persian and English). He is very motivated and interested in finding simple ways to teach complex topics and sharing them with others.

1 Cartesian Tensor Formalism and Dyadic Algebra

To derive, manipulate, simplify, and interpret various governing equations in continuum mechanics, we need a set of mathematical operators. These mathematical elements act like letters or words needed to write a poem. When we are writing or reading governing equations of continuum mechanics, it is customary to use two languages: vector notation and tensor algebra.

In vector notation, we define a symbol for each mathematical operation. It is easy to extend the vector operators to orthogonal curvilinear coordinates like spherical and polar coordinates. On the other hand, the difficulty of working with vector notation is the sophisticated use of symbols based on their definitions. Even we have to be worried about the validity of an operation in vector notation. For example, you need to define how to compute the divergence of a vector, the curl of a matrix, or the divergence of a scalar (which is undefined). You can easily obtain the derivative of the product of two functions, but it is not straightforward to calculate the divergence of the product of two matrices.

As an example of vector algebra operations, transposing is flipping the components of a matrix with respect to its diagonal pivot. Transpose of $n \times m$ matrix is $m \times n$ matrix. Transpose of a vertical vector ($n \times 1$) is a horizontal vector ($1 \times n$).

The key idea behind the tensor notation is to convert different vector operators into ordinary algebraic operations and derivatives. In vector notation, we need to present a specific definition for all operations such as the dot product, the double dot product, the dyadic product, and so forth. However, we can get rid of such numerous and sometimes vertiginous definitions by using the tensor notation rules. For example, instead of defining the divergence of a vector and the rank of its output (a scalar in this case), we prefer to use a simple algebraic relation that naturally generates a scalar.

One of well-known rules in tensor notation is the Einstein's summation convention which is used to simplify mathematical expressions. The best definition is presented by Einstein himself: "I have made a great discovery in mathematics. I have suppressed the summation sign every time that the summation must be made over an index that occurs twice".

Instead, if we intend to extend tensor notation to the non-Euclidean space or the curvilinear coordinates, we have to introduce the contravariant or upper indices, which makes the problem rather complicated. Fortunately, in continuum mechanics, we mostly do not deal with non-Euclidean spaces.[1] Hence, this chapter is just devoted to the cartesian tensor formalism, which is more straightforward for Euclidean

DOI: 10.1201/9781032719405-1

space. We will extend the vector notation for polar and spherical coordinates in Section 1.2.

1.1 VECTOR ALGEBRA

In this section, we are going to introduce the required symbols and definitions in vector notation often called the symbolic or the Gibbs notation. If the algebra related to the dyads is also included, the vector formalism is known as dyadic algebra. Important definitions commonly used in vector algebra are presented among the following items.

- **The material derivative**: In various physical sciences, we need to analyze changes. If the change of a quantity in time (the temporal rate) is needed, we have to use the material derivative operator (Section 2.4). If the spatial variation is under study, the gradient operator will be defined in the next item.
- **Polyads**, consisting of monads, dyads, triads, and tetrads, are mathematical quantities, which can be constructed using 1, 2, 3, and 4 vectors, respectively. These numbers are called the order (or sometimes the rank) of polyads. A vector (a first-order tensor) is a monad, a matrix (a second-order tensor) is a dyad, a third-order tensor is a triad, and so forth.
- **Trace** of a matrix or its first invariant is defined as the sum of its diagonal components.
- **Orthogonal matrix** has identical transpose and inverse with the determinant equal to 1 or -1.

$$\mathbf{A}^{-1} = \mathbf{A}^{T} \tag{1.1}$$

- **Rank of matrix** is the maximum order of all non-zero-determinant square sub-matrices. For example, the rank of the following matrix equals 1, since the determinant of all two by two square sub-matrices is zero.

$$\begin{pmatrix} 5 & -2 & 1 \\ -10 & 4 & -2 \end{pmatrix} \tag{1.2}$$

- **Transposing** is flipping the components of a matrix with respect to its diagonal pivot. So, transpose of $n \times m$ matrix is $m \times n$ matrix. Transpose of a vertical vector ($n \times 1$) is a horizontal vector ($1 \times n$).
- **Singular value decomposition**: You may decompose a non-square matrix (A) based on the singular value decomposition: $A = U\Sigma V^{T}$, where Σ is a diagonal matrix containing the square roots of eigenvalues of AA^{T} or $A^{T}A$ in descending order, U and V are the left and the right singular matrices with eigenvectors of AA^{T} and $A^{T}A$ as their columns, respectively. The number of non-zero singular values on the pivot of Σ equals the rank of A. SVD is the fundamental theorem of linear algebra that is important in low-rank matrix approximation, principal component analysis, dimensionality reduction, and unsupervised machine learning.

- **The dyadic product, the outer product, the tensor product, or the open product** uses two vectors to define a matrix. Hence, the dyadic product of two vectors produces a diad or a matrix, or a second-order tensor,

$$\mathbf{A} \otimes \mathbf{B} = \mathbf{A}\mathbf{B}^T = \begin{pmatrix} A_x \\ A_y \\ A_z \end{pmatrix} (B_x, B_y, B_z) = \begin{pmatrix} A_x B_x & A_x B_y & A_x B_z \\ A_y B_x & A_y B_y & A_y B_z \\ A_z B_x & A_z B_y & A_z B_z \end{pmatrix} \quad (1.3)$$

Consider that the outer product is not commutative. It means that $\mathbf{A} \otimes \mathbf{B} \neq \mathbf{B} \otimes \mathbf{A}$.

- **The contracted product or the dot product**: The dot product of two vectors $(\mathbf{A}, \mathbf{B})$, a vector and a matrix $(\mathbf{u}, \boldsymbol{\tau})$, and two matrices $(\mathbf{T}, \boldsymbol{\tau})$ respectively are defines as follows:

$$\mathbf{A} \cdot \mathbf{B} = \mathbf{A}^T \mathbf{B} = (A_x, A_y, A_z) \begin{pmatrix} B_x \\ B_y \\ B_z \end{pmatrix} = A_x B_x + A_y B_y + A_z B_z = trace(\mathbf{A} \otimes \mathbf{B})$$

$$\mathbf{u} \cdot \boldsymbol{\tau} = (\mathbf{u} \cdot first\ row, \mathbf{u} \cdot second\ row, \mathbf{u} \cdot third\ row)^T$$

$$= \begin{pmatrix} u\tau_{11} + v\tau_{12} + w\tau_{13} \\ u\tau_{21} + v\tau_{22} + w\tau_{23} \\ u\tau_{31} + v\tau_{32} + w\tau_{33} \end{pmatrix}$$

$$\mathbf{T} \cdot \boldsymbol{\tau} = \mathbf{T}\boldsymbol{\tau} =$$

$$\begin{pmatrix} T_{11}\tau_{11} + T_{12}\tau_{21} + T_{13}\tau_{31} & T_{11}\tau_{12} + T_{12}\tau_{22} + T_{13}\tau_{32} & T_{11}\tau_{13} + T_{12}\tau_{23} + T_{13}\tau_{33} \\ T_{21}\tau_{11} + T_{22}\tau_{21} + T_{23}\tau_{31} & T_{21}\tau_{12} + T_{22}\tau_{22} + T_{23}\tau_{32} & T_{21}\tau_{13} + T_{22}\tau_{23} + T_{23}\tau_{33} \\ T_{31}\tau_{11} + T_{32}\tau_{21} + T_{33}\tau_{31} & T_{31}\tau_{12} + T_{32}\tau_{22} + T_{33}\tau_{32} & T_{31}\tau_{13} + T_{32}\tau_{23} + T_{33}\tau_{33} \end{pmatrix}$$

$$(1.4)$$

where $\mathbf{u} = (u, v, w)$. As it is obvious, the contracted product of two matrices is the same as the common product of two matrices. Also, it is seen that the dot product of two vectors is equal to the contraction (trace) of the dyadic product of them.

- **The double contracted product, the double dot product, the inner scalar product, or the double divergence** of two matrices is defined as the contraction (trace) of the dot product of them,

$$\mathbf{T} : \boldsymbol{\tau} = trace(\mathbf{T} \cdot \boldsymbol{\tau}) = (T_{11}\tau_{11} + T_{12}\tau_{21} + T_{13}\tau_{31}) + (T_{21}\tau_{12} + T_{22}\tau_{22} + T_{23}\tau_{32}) + (T_{31}\tau_{13} + T_{32}\tau_{23} + T_{33}\tau_{33})$$

$$(1.5)$$

The double dot product of two matrices may be used to compute the second invariant of a tensor, the von Mises yield criterion in plasticity, modeling of non-Newtonian fluids, derivation of the dissipation term in the energy equation, and calculation of the Frobenius norm of a matrix. It will be proved that the double dot product of a symmetric tensor (such as the stress tensor in most applications) and an anti-symmetric tensor (such as the anti-symmetric part of the velocity gradient tensor) equals zero.

As you see, each dot product reduces the sum of the order of inputs by 2. For example, the dot product of two matrices is another matrix that is 2 orders less than the sum of the orders of two input matrices $(2 + 2 - 2 = 4 - 2 = 2)$. Or the double dot product of two matrices is 4 orders less than the sum of order of two input matrices, which is a scalar $(4 - 2 \times 2 = 0)$. As another example, the dot product of two vectors is a scalar $(1 + 1 - 2 = 0)$, and the dot product of a vector and a matrix is a vector $(1 + 2 - 2 = 1)$. This procedure is called contraction, which will be more deeply discussed using tensor notation.

· **The gradient operator**: The gradient vector is the fundamental operator of the differential calculus both in vector and tensor notations. The reason lies on the fact that in all physical sciences, we are working with changes in variables. The gradient operator shown by the nabla symbol and sometimes called the del vector formulates the spatial variation of a quantity.[2] If the gradient of a parameter in a direction is zero, it means that it has a uniform distribution in that direction.

For instance, if the x-component of the temperature gradient vector is the only non-zero component, it means that the temperature only varies along the x-direction. The gradient vector in cartesian coordinates is

$$\nabla = \left(\frac{\partial}{\partial x}, \frac{\partial}{\partial y}, \frac{\partial}{\partial z} \right)^T \tag{1.6}$$

where the superscript T means transpose and is used just for the sake of compact formulation. Based on the definition of the gradient vector, the total variation of a quantity like ϕ is defined as $d\phi = \nabla\phi \cdot d\mathbf{r} = (\partial\phi/\partial x)dx + (\partial\phi/\partial y)dy + (\partial\phi/\partial z)dz$.

The gradient operator always raises the order of the input quantities by one. It means that the gradient of a scalar (order 0) and a vector (order 1) are a vector (order 1) and a matrix (order 2), respectively. From the physical viewpoint, if someone asks about the spatial change of a scalar quantity (such as temperature), our answer: it is an incomplete question! To close the question, we need to have the direction of change of the quantity. In three-dimensional space, the question has three possible answers which are components of the temperature gradient vector,

$$\nabla T = \left(\frac{\partial T}{\partial x}, \frac{\partial T}{\partial y}, \frac{\partial T}{\partial z} \right)^T \tag{1.7}$$

One step forward, suppose that I ask about the variation of the velocity vector $(\mathbf{u} = (u, v, w))$. Again the question is ambiguous. I have to specify which component of the velocity vector and in which direction. In three-dimensional space, the question has nine possible answers. These nine answers form a matrix or a second-rank tensor $(\mathbf{L})$ as follows,

$$\mathbf{L} = \nabla\mathbf{u} = \begin{pmatrix} \frac{\partial u}{\partial x} & \frac{\partial u}{\partial y} & \frac{\partial u}{\partial z} \\ \frac{\partial v}{\partial x} & \frac{\partial v}{\partial y} & \frac{\partial v}{\partial z} \\ \frac{\partial w}{\partial x} & \frac{\partial w}{\partial y} & \frac{\partial w}{\partial z} \end{pmatrix} = \begin{pmatrix} \nabla u \\ \nabla v \\ \nabla w \end{pmatrix} = (\nabla \otimes \mathbf{u})^T \qquad (1.8)$$

The gradient of a vector equals the transpose of the dyadic product of the del operator and the vector.

The gradient of a scalar quantity appears in many conservation laws or constitutive relations such as the gradient of pressure in the momentum equation, the gradient of temperature in Fourier's law, and the gradient of concentration in the mass diffusion problem. Maybe the most common gradient-of-a-vector in continuum mechanics is the gradient of the velocity field, which plays a key role in the computation of the strain-rate tensor to obtain the momentum constitutive relations.

· **The Jacobian matrix**: The definition of the gradient of a vector that demonstrates the change of that vector in space is a starting point to present a definition for the derivative of two vectors with respect to each other. The derivative of a vector with respect to another vector is called the Jacobian matrix.[3]

The determinant of the Jacobian matrix (J) is used to investigate the uniqueness of transformations between two frames such as between the material (Lagrangian) and the spatial (Eulerian) frames (see Section 2.4). The determinant of the Jacobian matrix represents the ratio of the deformed and the initial volumes during a transformation. The Jacobian of a vector, like the deformation vector $(\mathbf{d})$, is defined as the gradient of the vector or the transpose of the dyadic product of the del operator by that vector,

$$\begin{aligned} \mathbf{J} &= \frac{d\mathbf{d}}{d\mathbf{x}} = \frac{d(d_x, d_y, d_z)}{d(x, y, z)} \\[2mm] &= \nabla\mathbf{d} = (\nabla \otimes \mathbf{d})^T = \begin{pmatrix} \frac{\partial d_x}{\partial x} & \frac{\partial d_x}{\partial y} & \frac{\partial d_x}{\partial z} \\ \frac{\partial d_y}{\partial x} & \frac{\partial d_y}{\partial y} & \frac{\partial d_y}{\partial z} \\ \frac{\partial d_z}{\partial x} & \frac{\partial d_z}{\partial y} & \frac{\partial d_z}{\partial z} \end{pmatrix} \qquad (1.9) \end{aligned}$$

· **The divergence operator** is defined as the dot product of the del operator and a vector or a matrix. If the divergence of a field is zero, it is called divergenceless, divergence-free, or solenoidal. The divergence operator always reduces the order of the input quantity by 1. So the minimum order of the input entry of the divergence operator shall be 1. As a result, the divergence of a scalar is meaningless.

The divergences of a vector and a tensor appear during the derivation of the conservation of mass/energy and momentum, respectively. The divergence of a vector is equivalent to the trace of the gradient of that vector which is a scalar. In cartesian coordinates

$$\nabla \cdot \mathbf{u} = trace(\nabla \mathbf{u}) = \frac{\partial u}{\partial x} + \frac{\partial v}{\partial y} + \frac{\partial w}{\partial z} \qquad (1.10)$$

where *trace* is the sum of diagonal components of the matrix. The divergence of a matrix is a vector whose components equal the divergence of the rows of that matrix,

$$\nabla \cdot \boldsymbol{\tau} = [\nabla \cdot (first\ row), \nabla \cdot (second\ row), \nabla \cdot (third\ row)]^T$$
$$= \begin{pmatrix} \frac{\partial \tau_{11}}{\partial x} + \frac{\partial \tau_{12}}{\partial y} + \frac{\partial \tau_{13}}{\partial z} \\ \frac{\partial \tau_{21}}{\partial x} + \frac{\partial \tau_{22}}{\partial y} + \frac{\partial \tau_{23}}{\partial z} \\ \frac{\partial \tau_{31}}{\partial x} + \frac{\partial \tau_{32}}{\partial y} + \frac{\partial \tau_{33}}{\partial z} \end{pmatrix} \qquad (1.11)$$

· **The curl operator**: The curl of a vector sometimes shown by rot() is defined as the cross product of the del operator and a vector. Curls of a matrix and a scalar appear in the definitions of compatibility conditions of elasticity and integral theorems. The curl of velocity equals vorticity, which is the most important quantity in vorticity dynamics. Since the del operator takes the derivative of its entry, it is not possible to find a general relation between orthogonality of a vector and its curl. In cartesian coordinates,

$$\nabla \times \mathbf{u} = det \begin{pmatrix} \mathbf{i} & \mathbf{j} & \mathbf{k} \\ \frac{\partial}{\partial x} & \frac{\partial}{\partial y} & \frac{\partial}{\partial z} \\ u & v & w \end{pmatrix} \qquad (1.12)$$

The definition of the curl operator stems from the Hodge-duality in topology. In n-dimensional space, the maximum order of the input of the curl operator is n−1. So, the curl of a matrix is meaningful just in three-dimensional space. The output of the curl of a quantity with order r in n-dimensional space is a quantity with order n-r-1. For example, in two-dimensional space (n=2), the curl of a vector (r=1) is a scalar,[4] and the curl of a scalar (r=0) is a vector. Similarly, in three-dimensional space, the curl of a matrix is a scalar, the curl of a vector is a vector, and the curl of a scalar is a matrix.

· **The Laplacian operator**: It will be proved that the divergence of the curl of any vector is zero. But, the divergence of the gradient of a quantity is called the Laplacian operator (Δ or ∇^2). Since the divergence operator reduces, and the gradient operator enhances the order of their input, the Laplacian operator preserves the order of the input function. It is easy to show that the divergence of the gradient of a scalar in cartesian coordinates equals

$$\nabla^2 \phi = \nabla \cdot \nabla \phi = \nabla \cdot \left(\frac{\partial \phi}{\partial x}, \frac{\partial \phi}{\partial y}, \frac{\partial \phi}{\partial z} \right) = \frac{\partial^2 \phi}{\partial x^2} + \frac{\partial^2 \phi}{\partial y^2} + \frac{\partial^2 \phi}{\partial z^2} \qquad (1.13)$$

The Laplacian of a vector in cartesian coordinates is a vector with the components equal to the Laplacian of each component of the vector. In curvilinear coordinates, the components of the Laplacian of a vector are not the same as the Laplacian of the components of that vector (for more details see Section 1.2).

- **The bilaplacian or the biharmonic operator**: When the Laplacian operator acts twice, a new operator is obtained, which is called the biharmonic operator,

$$\nabla^4 \phi = \nabla^2 \nabla^2 \phi = \Delta^2 \phi \qquad (1.14)$$

It is easy to prove that the bilaplacian operator in cartesian coordinates is

$$
\begin{aligned}
\nabla^4 &= \frac{\partial^2}{\partial x^2}\left(\frac{\partial^2}{\partial x^2}+\frac{\partial^2}{\partial y^2}+\frac{\partial^2}{\partial z^2}\right)+\frac{\partial^2}{\partial y^2}\left(\frac{\partial^2}{\partial x^2}+\frac{\partial^2}{\partial y^2}+\frac{\partial^2}{\partial z^2}\right) \\
&+ \frac{\partial^2}{\partial z^2}\left(\frac{\partial^2}{\partial x^2}+\frac{\partial^2}{\partial y^2}+\frac{\partial^2}{\partial z^2}\right) \\
&= \frac{\partial^4}{\partial x^4}+\frac{\partial^4}{\partial y^4}+\frac{\partial^4}{\partial z^4}+2\frac{\partial^4}{\partial x^2\partial y^2}+2\frac{\partial^4}{\partial x^2\partial z^2}+2\frac{\partial^4}{\partial y^2\partial z^2} \quad (1.15)
\end{aligned}
$$

If the Laplacian or the bilaplacian of a quantity equals zero, the yielded equation is called the Laplace equation or the biharmonic equation, respectively. If the Laplacian of a parameter equals a non-zero function, the equation is called the Poisson equation.

This operator appears in the equation of motion of the linearized creeping flows, the theory of elasticity, and the Kuramoto–Sivashinsky equation for the investigation of diffusive instabilities in a laminar flame front.

1.2 CURVILINEAR ORTHOGONAL COORDINATES

The vectorial form of the operators in cylindrical and spherical coordinates will be derived in this section. Complexities in formulae arise from the existence of partial derivatives of the unit vectors with respect to different directions. There are three coordinates and three unit vectors; so we can define nine partial derivatives.

In cylindrical coordinates (r, θ, z),

$$
\begin{aligned}
\mathbf{e}_r &= \mathbf{i}\cos\theta +\mathbf{j}\sin\theta \\
\mathbf{e}_\theta &= -\mathbf{i}\sin\theta +\mathbf{j}\cos\theta \\
\mathbf{e}_z &= \mathbf{k}
\end{aligned} \qquad (1.16)
$$

All partial derivatives are zero except two of them,

$$\frac{\partial \mathbf{e}_r}{\partial \theta} = -\mathbf{i}\sin\theta +\mathbf{j}\cos\theta = \mathbf{e}_\theta, \qquad \frac{\partial \mathbf{e}_\theta}{\partial \theta} = -\mathbf{i}\cos\theta -\mathbf{j}\sin\theta = -\mathbf{e}_r \qquad (1.17)$$

In spherical coordinates (R, ϕ, θ),

$$
\begin{aligned}
\mathbf{e}_R &= \mathbf{i}\sin\theta\cos\phi +\mathbf{j}\sin\theta\sin\phi +\mathbf{k}\cos\theta \\
\mathbf{e}_\phi &= -\mathbf{i}\sin\phi +\mathbf{j}\cos\phi \\
\mathbf{e}_\theta &= \mathbf{i}\cos\theta\cos\phi +\mathbf{j}\cos\theta\sin\phi -\mathbf{k}\sin\theta
\end{aligned} \qquad (1.18)
$$

where θ is the polar angle varying between 0 and π, and ϕ is the azimuthal angle varying between 0 and 2π. We can find five non-zero partial derivatives of the unit vectors,

$$\frac{\partial \mathbf{e}_R}{\partial \theta} = \mathbf{e}_\theta, \quad \frac{\partial \mathbf{e}_R}{\partial \phi} = \mathbf{e}_\phi \sin \theta, \quad \frac{\partial \mathbf{e}_\theta}{\partial \phi} = \mathbf{e}_\phi \cos \theta,$$

$$\frac{\partial \mathbf{e}_\phi}{\partial \phi} = -(\mathbf{e}_R \sin \theta + \mathbf{e}_\theta \cos \theta), \quad \frac{\partial \mathbf{e}_\theta}{\partial \theta} = -\mathbf{e}_R \qquad (1.19)$$

Example 1: Obtain the differential line element vector in spherical coordinates.

$$\begin{aligned}
d\mathbf{R} &= d(R\mathbf{e}_R) = \mathbf{e}_R dR + R d\mathbf{e}_R = \mathbf{e}_R dR + R \left(\underbrace{\frac{\partial \mathbf{e}_R}{\partial R}}_{0} dR + \frac{\partial \mathbf{e}_R}{\partial \theta} d\theta + \frac{\partial \mathbf{e}_R}{\partial \phi} d\phi \right) \\
&= dR\mathbf{e}_R + Rd\theta\mathbf{e}_\theta + R \sin \theta d\phi \mathbf{e}_\phi \qquad (1.20)
\end{aligned}$$

In general, the coefficients of the unit vectors in the line element vector play a key role in the derivation of formulae in various coordinates. To obtain different operators in various coordinates you just need to compute these factors based on geometrical relations. In this case, the coefficients are $h_1 = 1$, $h_2 = R$, and $h_3 = R \sin \theta$. For cylindrical coordinates, these scale factors equal $h_1 = 1$, $h_2 = r$, $h_3 = 1$. For cartesian coordinate, h_2 also equals 1.

Example 2: Obtain the gradient operator in spherical coordinates.

Based on the definition of the gradient vector $df = \nabla f \cdot d\mathbf{R}$ and the differential area derived in Example 1, the del operator in spherical coordinates is

$$\begin{aligned}
df &= \frac{\partial f}{\partial R} dR + \frac{\partial f}{\partial \theta} d\theta + \frac{\partial f}{\partial \phi} d\phi \\
\nabla f \cdot d\mathbf{R} &= (\nabla f)_R dR + (\nabla f)_\theta R d\theta + (\nabla f)_\phi R \sin \theta d\phi \rightarrow \\
(\nabla f)_R &= \frac{\partial f}{\partial R}, \quad (\nabla f)_\theta = \frac{1}{R} \frac{\partial f}{\partial \theta}, \quad (\nabla f)_\phi = \frac{1}{R \sin \theta} \frac{\partial f}{\partial \phi} \qquad (1.21)
\end{aligned}$$

Generally speaking, the coefficients of components of the gradient operator are inverse of scale factors: $\frac{1}{h_1} = 1$, $\frac{1}{h_2} = \frac{1}{R}$, $\frac{1}{h_3} = \frac{1}{R \sin \theta}$.

Example 3: Obtain the divergence of a vector in cylindrical coordinates.

Using the definition of the divergence in symbolic form

$$\begin{aligned}
\nabla \cdot \mathbf{A} &= \left(\mathbf{e}_r \frac{\partial}{\partial r} + \mathbf{e}_\theta \frac{1}{r} \frac{\partial}{\partial \theta} + \mathbf{e}_z \frac{\partial}{\partial z} \right) \cdot \mathbf{A} \\
&= \mathbf{e}_r \cdot \frac{\partial \mathbf{A}}{\partial r} + \frac{\mathbf{e}_\theta}{r} \cdot \frac{\partial \mathbf{A}}{\partial \theta} + \mathbf{e}_z \cdot \frac{\partial \mathbf{A}}{\partial z} \qquad (1.22)
\end{aligned}$$

Consider that the derivative is prior to the dot product. Using the rule for the derivative of product and $\mathbf{A} = A_r\mathbf{e}_r + A_\theta\mathbf{e}_\theta + A_z\mathbf{e}_z$,

$$
\begin{aligned}
\nabla \cdot \mathbf{A} \;=\;& \mathbf{e}_r \cdot \left(\frac{\partial A_r}{\partial r}\mathbf{e}_r + \frac{\partial A_\theta}{\partial r}\mathbf{e}_\theta + \frac{\partial A_z}{\partial r}\mathbf{e}_z + A_r\frac{\partial \mathbf{e}_r}{\partial r} + A_\theta\frac{\partial \mathbf{e}_\theta}{\partial r} + A_z\frac{\partial \mathbf{e}_z}{\partial r} \right) \\
+\;& \frac{\mathbf{e}_\theta}{r} \cdot \left(\frac{\partial A_r}{\partial \theta}\mathbf{e}_r + \frac{\partial A_\theta}{\partial \theta}\mathbf{e}_\theta + \frac{\partial A_z}{\partial \theta}\mathbf{e}_z + A_r\frac{\partial \mathbf{e}_r}{\partial \theta} + A_\theta\frac{\partial \mathbf{e}_\theta}{\partial \theta} + A_z\frac{\partial \mathbf{e}_z}{\partial \theta} \right) \\
+\;& \mathbf{e}_z \cdot \left(\frac{\partial A_r}{\partial z}\mathbf{e}_r + \frac{\partial A_\theta}{\partial z}\mathbf{e}_\theta + \frac{\partial A_z}{\partial z}\mathbf{e}_z + A_r\frac{\partial \mathbf{e}_r}{\partial z} + A_\theta\frac{\partial \mathbf{e}_\theta}{\partial z} + A_z\frac{\partial \mathbf{e}_z}{\partial z} \right) \rightarrow \\
\nabla \cdot \mathbf{A} \;=\;& \mathbf{e}_r \cdot \left(\frac{\partial A_r}{\partial r}\mathbf{e}_r + \frac{\partial A_\theta}{\partial r}\mathbf{e}_\theta + \frac{\partial A_z}{\partial r}\mathbf{e}_z + 0 + 0 + 0 \right) \\
+\;& \frac{\mathbf{e}_\theta}{r} \cdot \left(\frac{\partial A_r}{\partial \theta}\mathbf{e}_r + \frac{\partial A_\theta}{\partial \theta}\mathbf{e}_\theta + \frac{\partial A_z}{\partial \theta}\mathbf{e}_z + A_r\mathbf{e}_\theta - A_\theta\mathbf{e}_r + 0 \right) \\
+\;& \mathbf{e}_z \cdot \left(\frac{\partial A_r}{\partial z}\mathbf{e}_r + \frac{\partial A_\theta}{\partial z}\mathbf{e}_\theta + \frac{\partial A_z}{\partial z}\mathbf{e}_z + 0 + 0 + 0 \right) \rightarrow \\
\nabla \cdot \mathbf{A} \;=\;& \frac{\partial A_r}{\partial r} + \frac{1}{r}\frac{\partial A_\theta}{\partial \theta} + \frac{A_r}{r} + \frac{\partial A_z}{\partial z} \\
=\;& \frac{1}{r}\frac{\partial (rA_r)}{\partial r} + \frac{1}{r}\frac{\partial A_\theta}{\partial \theta} + \frac{\partial A_z}{\partial z}
\end{aligned}
\tag{1.23}
$$

Generally, the divergence operator in other curvilinear coordinates with unit vectors x_1, x_2, x_3 equals $\frac{1}{h_1 h_2 h_3}\left(\frac{\partial(h_2 h_3 A_1)}{\partial x_1} + \frac{\partial(h_1 h_3 A_2)}{\partial x_2} + \frac{\partial(h_1 h_2 A_3)}{\partial x_3} \right)$.

The following operators in both cylindrical and spherical coordinates can be derived similarly. The first and the second equations in each item correspond to the cylindrical and spherical coordinates, respectively.

· **Differential line element**

$$
\begin{aligned}
d\mathbf{R} &= dr\mathbf{e}_r + rd\theta\mathbf{e}_\theta + dz\mathbf{e}_z \\
d\mathbf{R} &= dR\mathbf{e}_R + Rd\theta\mathbf{e}_\theta + R\sin\theta d\phi\mathbf{e}_\phi
\end{aligned}
\tag{1.24}
$$

· **Differential surface element**

$$
\begin{aligned}
d\mathbf{S} &= rd\theta dz\mathbf{e}_r + drdz\mathbf{e}_\theta + rdrd\theta\mathbf{e}_z \\
d\mathbf{S} &= R^2\sin\theta d\theta d\phi\mathbf{e}_R + R\sin\theta dRd\phi\mathbf{e}_\theta + RdRd\theta\mathbf{e}_\phi
\end{aligned}
\tag{1.25}
$$

- **Differential volume** can be obtained using the determinant of the Jacobian matrix,

$$
\begin{aligned}
d\forall &= det\begin{pmatrix} \frac{\partial x}{\partial r} & \frac{\partial x}{\partial \theta} & \frac{\partial x}{\partial z} \\ \frac{\partial y}{\partial r} & \frac{\partial y}{\partial \theta} & \frac{\partial y}{\partial z} \\ \frac{\partial z}{\partial r} & \frac{\partial z}{\partial \theta} & \frac{\partial z}{\partial z} \end{pmatrix} dr\,d\theta\,dz \\
&= det\begin{pmatrix} \cos\theta & -r\sin\theta & 0 \\ \sin\theta & r\cos\theta & 0 \\ 0 & 0 & 1 \end{pmatrix} dr\,d\theta\,dz = r\,dr\,d\theta\,dz \\
d\forall &= R^2\sin\theta\,dR\,d\theta\,d\phi
\end{aligned}
\tag{1.26}
$$

- **Differential solid angle** is defined just in spherical coordinates

$$
d\Omega = \frac{dS_R}{R^2} = \sin\theta\,d\theta\,d\phi
\tag{1.27}
$$

- **Gradient of a scalar**

$$
\begin{aligned}
\nabla f &= \frac{\partial f}{\partial r}\mathbf{e}_r + \frac{1}{r}\frac{\partial f}{\partial \theta}\mathbf{e}_\theta + \frac{\partial f}{\partial z}\theta\mathbf{e}_z \\
\nabla f &= \frac{\partial f}{\partial R}\mathbf{e}_R + \frac{1}{R}\frac{\partial f}{\partial \theta}\mathbf{e}_\theta + \frac{1}{R\sin\theta}\frac{\partial f}{\partial \phi}\mathbf{e}_\phi
\end{aligned}
\tag{1.28}
$$

- **Divergence of a vector**

$$
\begin{aligned}
\nabla\cdot\mathbf{A} &= \frac{1}{r}\frac{\partial(rA_r)}{\partial r} + \frac{1}{r}\frac{\partial A_\theta}{\partial \theta} + \frac{\partial A_z}{\partial z} \\
\nabla\cdot\mathbf{A} &= \frac{1}{R^2}\frac{\partial(R^2 A_R)}{\partial R} + \frac{1}{R\sin\theta}\frac{\partial(A_\theta\sin\theta)}{\partial \theta} + \frac{1}{R\sin\theta}\frac{\partial A_\phi}{\partial \phi}
\end{aligned}
\tag{1.29}
$$

- **Curl of a vector:** The first component of the curl vector can be expressed based on the scale factors: $\frac{1}{h_2 h_3}\left(\frac{\partial(h_3 A_3)}{\partial x_2} - \frac{\partial(h_2 A_2)}{\partial x_3}\right)$. Other components are similar to the first component.

$$
\begin{aligned}
\nabla\times\mathbf{A} &= \begin{vmatrix} \mathbf{e}_r & \mathbf{e}_\theta & \mathbf{e}_z \\ \frac{\partial}{\partial r} & \frac{\partial}{\partial \theta} & \frac{\partial}{\partial z} \\ A_r & rA_\theta & A_z \end{vmatrix} \\
&= \left(\frac{1}{r}\frac{\partial A_z}{\partial \theta} - \frac{\partial A_\theta}{\partial z}\right)\mathbf{e}_r + \left(\frac{\partial A_r}{\partial z} - \frac{\partial A_z}{\partial r}\right)\mathbf{e}_\theta + \frac{1}{r}\left(\frac{\partial(rA_\theta)}{\partial r} - \frac{\partial A_r}{\partial \theta}\right)\mathbf{e}_z \\
\nabla\times\mathbf{A} &= \begin{vmatrix} \frac{\mathbf{e}_R}{R^2\sin\theta} & \frac{\mathbf{e}_\theta}{R\sin\theta} & \frac{\mathbf{e}_\phi}{R} \\ \frac{\partial}{\partial R} & \frac{\partial}{\partial \theta} & \frac{\partial}{\partial \phi} \\ A_R & RA_\theta & R\sin\theta A_\phi \end{vmatrix} \\
&= \frac{1}{R\sin\theta}\left(\frac{\partial(A_\phi\sin\theta)}{\partial \theta} - \frac{\partial A_\theta}{\partial \phi}\right)\mathbf{e}_R + \frac{1}{R}\left(\frac{1}{\sin\theta}\frac{\partial A_R}{\partial \phi} - \frac{\partial(RA_\phi)}{\partial R}\right)\mathbf{e}_\theta \\
&\quad + \frac{1}{R}\left(\frac{\partial(RA_\theta)}{\partial R} - \frac{\partial A_R}{\partial \theta}\right)\mathbf{e}_\phi
\end{aligned}
\tag{1.30}
$$

- **Laplacian of a scalar**

$$\nabla^2 f = \frac{1}{r}\frac{\partial}{\partial r}\left(r\frac{\partial f}{\partial r}\right) + \frac{1}{r^2}\frac{\partial^2 f}{\partial \theta^2} + \frac{\partial^2 f}{\partial z^2}$$

$$\nabla^2 f = \frac{1}{R^2}\frac{\partial}{\partial R}\left(R^2\frac{\partial f}{\partial R}\right) + \frac{1}{R^2\sin\theta}\frac{\partial}{\partial\theta}\left(\sin\theta\frac{\partial f}{\partial\theta}\right) + \frac{1}{R^2\sin^2\theta}\frac{\partial^2 f}{\partial\phi^2}$$

$$(1.31)$$

- **Laplacian of a vector**

$$\nabla^2\mathbf{A} = \left(\nabla^2 A_r - \frac{A_r}{r^2} - \frac{2}{r^2}\frac{\partial A_\theta}{\partial\theta}\right)\mathbf{e}_r + \left(\nabla^2 A_\theta - \frac{A_\theta}{r^2} + \frac{2}{r^2}\frac{\partial A_r}{\partial\theta}\right)\mathbf{e}_\theta + \nabla^2 A_z\mathbf{e}_z$$

$$\nabla^2\mathbf{A} = \left(\nabla^2 A_R - \frac{2A_R}{R^2} - \frac{2}{R^2\sin\theta}\frac{\partial(A_\theta\sin\theta)}{\partial\theta} - \frac{2}{R^2\sin\theta}\frac{\partial A_\phi}{\partial\phi}\right)\mathbf{e}_R$$

$$+ \left(\nabla^2 A_\theta - \frac{A_\theta}{R^2\sin^2\theta} + \frac{2}{R^2}\frac{\partial A_R}{\partial\theta} - \frac{2\cos\theta}{R^2\sin^2\theta}\frac{\partial A_\phi}{\partial\phi}\right)\mathbf{e}_\theta$$

$$+ \left(\nabla^2 A_\phi - \frac{A_\phi}{R^2\sin^2\theta} + \frac{2}{R^2\sin^2\theta}\frac{\partial A_R}{\partial\phi} + \frac{2\cos\theta}{R^2\sin^2\theta}\frac{\partial A_\theta}{\partial\phi}\right)\mathbf{e}_\phi \quad (1.32)$$

- **Divergence of a tensor**

$$\nabla\cdot\mathbf{T} = \left(\frac{\partial T_{rr}}{\partial r} + \frac{1}{r}\frac{\partial T_{\theta r}}{\partial\theta} + \frac{\partial T_{zr}}{\partial z} + \frac{1}{r}(T_{rr} - T_{\theta\theta})\right)\mathbf{e}_r$$

$$+ \left(\frac{\partial T_{r\theta}}{\partial r} + \frac{1}{r}\frac{\partial T_{\theta\theta}}{\partial\theta} + \frac{\partial T_{z\theta}}{\partial z} + \frac{1}{r}(T_{r\theta} + T_{\theta r})\right)\mathbf{e}_\theta$$

$$+ \left(\frac{\partial T_{rz}}{\partial r} + \frac{1}{r}\frac{\partial T_{\theta z}}{\partial\theta} + \frac{\partial T_{zz}}{\partial z} + \frac{1}{r}T_{rz}\right)\mathbf{e}_z \qquad (1.33)$$

$$\nabla\cdot\mathbf{T} = \left(\frac{\partial T_{RR}}{\partial R} + 2\frac{T_{RR}}{R} + \frac{1}{R}\frac{\partial T_{\theta r}}{\partial\theta} + \frac{\cot\theta}{R}T_{\theta R} + \frac{1}{R\sin\theta}\frac{\partial T_{\phi R}}{\partial\phi} - \frac{1}{R}(T_{\theta\theta} + T_{\phi\phi})\right)\mathbf{e}_R$$

$$+ \left(\frac{\partial T_{R\theta}}{\partial R} + 2\frac{T_{R\theta}}{R} + \frac{1}{R}\frac{\partial T_{\theta\theta}}{\partial\theta} + \frac{\cot\theta}{R}T_{\theta\theta} + \frac{1}{R\sin\theta}\frac{\partial T_{\phi\theta}}{\partial\phi} + \frac{T_{\theta R}}{R} - \frac{\cot\theta}{R}T_{\phi\phi}\right)\mathbf{e}_\theta$$

$$+ \left(\frac{\partial T_{R\phi}}{\partial R} + 2\frac{T_{R\phi}}{R} + \frac{1}{R}\frac{\partial T_{\theta\phi}}{\partial\theta} + \frac{1}{R\sin\theta}\frac{\partial T_{\phi\phi}}{\partial\phi} + \frac{T_{\phi R}}{R} + \frac{\cot\theta}{R}(T_{\phi\theta} + T_{\theta\phi})\right)\mathbf{e}_\phi$$

$$(1.34)$$

- **Biharmonic operator**

$$\nabla^4 f = \frac{1}{r}\frac{\partial}{\partial r}\left(r\frac{\partial}{\partial r}\left(\frac{1}{r}\frac{\partial}{\partial r}\left(r\frac{\partial f}{\partial r}\right)\right)\right)$$

$$+ \frac{2}{r^2}\frac{\partial^4 f}{\partial\theta^2\partial r^2} + \frac{1}{r^4}\frac{\partial^4 f}{\partial\theta^4} - \frac{2}{r^3}\frac{\partial^3 f}{\partial\theta^2\partial r} + \frac{4}{r^4}\frac{\partial^2 f}{\partial\theta^2}$$

$$\nabla^4 f = \left[\frac{1}{R^2}\frac{\partial}{\partial R}\left(R^2\frac{\partial f}{\partial R}\right) + \frac{1}{R^2\sin\theta}\frac{\partial}{\partial\theta}\left(\sin\theta\frac{\partial f}{\partial\theta}\right)\right]^2 \quad (1.35)$$

- **Material derivative** will be discussed in Section 2.4,

$$
\frac{D\mathbf{B}}{Dt} = \frac{\partial \mathbf{B}}{\partial t} \; + \; \left(u_r \frac{\partial B_r}{\partial r} + \frac{u_\theta}{r} \frac{\partial B_r}{\partial \theta} + u_z \frac{\partial B_r}{\partial z} - \frac{u_\theta B_\theta}{r} \right) \mathbf{e}_r
$$
$$
+ \; \left(u_r \frac{\partial B_\theta}{\partial r} + \frac{u_\theta}{r} \frac{\partial B_\theta}{\partial \theta} + u_z \frac{\partial B_\theta}{\partial z} + \frac{u_\theta B_r}{r} \right) \mathbf{e}_\theta
$$
$$
+ \; \left(u_r \frac{\partial B_z}{\partial r} + \frac{u_\theta}{r} \frac{\partial B_z}{\partial \theta} + u_z \frac{\partial B_z}{\partial z} \right) \mathbf{e}_z \qquad (1.36)
$$

$$
\frac{D\mathbf{B}}{Dt} = \frac{\partial \mathbf{B}}{\partial t} + \left(u_R \frac{\partial B_R}{\partial R} + \frac{u_\theta}{R} \frac{\partial B_R}{\partial \theta} + \frac{u_\phi}{R\sin\theta} \frac{\partial B_R}{\partial \phi} - \frac{u_\theta B_\theta + u_\phi B_\phi}{R} \right) \mathbf{e}_R
$$
$$
+ \left(u_R \frac{\partial B_\theta}{\partial R} + \frac{u_\theta}{R} \frac{\partial B_\theta}{\partial \theta} + \frac{u_\phi}{R\sin\theta} \frac{\partial B_\theta}{\partial \phi} + \frac{u_\theta B_R}{R} - \frac{u_\phi B_\phi \cot\theta}{R} \right) \mathbf{e}_\theta
$$
$$
+ \left(u_R \frac{\partial B_\phi}{\partial R} + \frac{u_\theta}{R} \frac{\partial B_\phi}{\partial \theta} + \frac{u_\phi}{R\sin\theta} \frac{\partial B_\phi}{\partial \phi} + \frac{u_\phi B_R}{R} + \frac{u_\phi B_\theta \cot\theta}{R} \right) \mathbf{e}_\phi
$$
$$
(1.37)
$$

- **The strain-rate tensor** (symmetric) will be introduced in Section 3.1.

$$
\begin{pmatrix}
\frac{\partial u_r}{\partial r} & \frac{1}{2}\left(\frac{1}{r}\frac{\partial u_r}{\partial \theta} + \frac{\partial u_\theta}{\partial r} - \frac{u_\theta}{r} \right) & \frac{1}{2}\left(\frac{\partial u_r}{\partial z} + \frac{\partial u_z}{\partial r} \right) \\
\cdot & \frac{1}{r}\frac{\partial u_\theta}{\partial \theta} + \frac{u_r}{r} & \frac{1}{2}\left(\frac{1}{r}\frac{\partial u_z}{\partial \theta} + \frac{\partial u_\theta}{\partial z} \right) \\
\cdot & \cdot & \frac{\partial u_z}{\partial z}
\end{pmatrix} \qquad (1.38)
$$

$$
\begin{pmatrix}
\frac{\partial u_R}{\partial R} & \frac{1}{2}\left[R\frac{\partial}{\partial R}\left(\frac{u_\theta}{R} \right) + \frac{1}{R}\frac{\partial u_R}{\partial \theta} \right] & \frac{1}{2}\left[\frac{1}{R\sin\theta}\frac{\partial u_R}{\partial \phi} + R\frac{\partial}{\partial R}\left(\frac{u_\phi}{R} \right) \right] \\
\cdot & \frac{1}{R}\frac{\partial u_\theta}{\partial \theta} + \frac{u_R}{R} & \frac{1}{2}\left[\frac{\sin\theta}{R}\frac{\partial}{\partial \theta}\left(\frac{u_\phi}{\sin\theta} \right) + \frac{1}{R\sin\theta}\frac{\partial u_\theta}{\partial \phi} \right] \\
\cdot & \cdot & \frac{1}{R\sin\theta}\frac{\partial u_\phi}{\partial \phi} + \frac{u_R}{R} + \frac{u_\theta \cot\theta}{R}
\end{pmatrix}
$$
$$
(1.39)
$$

Other curvilinear coordinates are the bipolar, parabolic, and elliptic two-dimensional coordinates and the toroidal, paraboloidal, ellipsoidal, prolate and oblate spheroidal three-dimensional coordinates.

1.3 CARTESIAN TENSOR CALCULUS

Our final goal in this section is to transform different symbols and definitions in vector notation into a new language called index notation, tensor notation, or Einstein's notation in which we just deal with simple algebraic operations. So, in tensor formalism, we do not need to present new definitions for new symbols like what we did for the dyadic product, the curl, or the divergence of a vector, dyads, or triads.

From a historical viewpoint, the definition of tensors goes back to Carl Friedrich Gauss, William Hamilton, Tullio Levi-Civita, Gregorio Ricci-Curbastro, and Albert Einstein.[5] They proposed using an index for each coordinate direction instead of working with polyads and numerous definitions of operations and symbols. For instance, a vector **u** is shown by a scaler u_i with an index i varying from 1 to 3 in three-dimensional space. Consequently, the gradient of a vector can be defined as $\frac{\partial u_i}{\partial x_j}$ containing two indices i and j both varying from 1 to 3, which produces nine components.

We need two indices to illustrate dyads in tensor notation. A dyad (a matrix) τ in tensor notation is replaced by a scalar τ_{ij} with two indices. Consequently, the gradient of $\tau_{ij} = \frac{\partial \tau_{ij}}{\partial x_k}$ contains three indices; hence, it is a triad or a third-order tensor.

The definition of tensor is the generalization of the concept of scalars, vectors, dyads, triads, or polyads to help readers easily work with these mathematical elements without being engaged with lots of definitions. For instance, in vector notation, you need to remember that the curl of a scalar in three-dimensional space is a dyad and vice versa. But in tensor notation, you just write the definition of the curl operator and you will implicitly find out what the order of the result is.

In some fields of physics which are engaged with non-Euclidean space such as the relativity theorem, or when we are working with curvilinear coordinates like the spherical and cylindrical coordinates, we need to define the contravariant notation. In the scope of this book, we just deal with Euclidean space. For the curvilinear coordinates, we will still use the vector notation rules.

The following items describe the fundamental definition of concepts and rules in tensor notation.

1. **Coordinate transformation rules:** Tensors must follow coordinate transformation laws. From the geometric viewpoint, any quantity which satisfies the coordinate transformation laws under coordinate change is a tensor. For example, we want to calculate the transformed version of a vector in another coordinate system. The vector is invariant in any transformed coordinates, although its components have changed. So, it can be a tensor. Similarly, density at a point in space has nothing to do with the reference coordinate system. It is unchanged, so scalars are also tensors. Again, suppose a moving fluid in a pipe. The stress state at a point in the flow should not be a function of the coordinate system. So, the stress tensor is a tensor!

2. **Rank/order/degree of a tensor** is the number of indices needed to define a tensor. The rank of a tensor is independent of the dimension of space. Based on this definition, the scalar is a zeroth-order tensor, the vector is a first-order tensor, the dyad ($m \times m$ matrix in m-dimensional space) is a second-order tensor, the triad is a third-order tensor, and the polyad is a general tensor of an arbitrary rank. For example, the gradient of a second-order tensor is a third-rank tensor, because needs three indices to be defined.

3. **The nth-rank tensors** represent different physical quantities similar to what polyads do in symbolic notation. Tensors are the fundamental alphabets of

the mathematical language of continuum mechanics, fluid dynamics, elasticity, electrodynamics, topology, and the general relativity theorem. Fundamentally, tensors are algebraic physical facts that are independent of coordinates and should obey specific transformation laws.[6]

The number of components needed to present an nth-order tensor in m-dimensional space is m^n. For instance, the number of components of a scalar, a vector, and a 3×3 matrix in three-dimensional space is 1, 3, and 9, respectively. The electromagnetic tensor is a second-rank four-dimensional tensor with 16 components, and the elasticity tensor is a fourth-rank tensor with 81 components in three-dimensional space.

4. **The rule of ranks:** Suppose that we have an unknown mathematical machine that accepts two vectors and produces another vector. In this example, the sum of the ranks of two inputs is 1+1=2 and the rank of the output is 1. The order of the unknown operator should be the sum of ranks of the inputs and the outputs (2+1=3 in this case).

Conversely, if an operator with three indices is applied to two vectors, the result should be another vector. The mentioned operator in this example could be the curl operator (or the third-order permutation symbol) that accepts two vectors (the del operator and an arbitrary vector) and generates another vector. Based on this rule, the result of the curl (order=3) of a scalar (order=0) should be a second-order tensor (3-(1+0)=2). Do not forget the order of the nabla vector which is another input of the curl operator. Similarly, the curl of a second-order tensor (rank=2) should be a scalar (3-(1+2)=0). In all these cases, the sum of the ranks of the input and the output arguments is 3.

We know that the matrix product of a 3×3 matrix and a 3×1 vector is a 3×1 vector. Let's reexpress this statement in tensor notation. The order of a tensor operator which accepts a vector and delivers another vector should be the sum of ranks of arguments: 1+1=2. So, the rank of the mapping operator in this case should be 2 (a second-order tensor or a 3×3 matrix). The elasticity tensor accepts a second-order tensor and delivers another second-order tensor. Hence based on the rule of ranks, the rank of the elasticity tensor should be 4.

5. **Mixed-type tensors versus cartesian tensors:** A tensor can be of contravariant, covariant, or mixed type. In the present book, the covariant tensors only containing the lower indices will be used like τ_{ij}. Upper indices appear in contravariant or mixed-type tensors like τ^i_{jk} which contains one upper and two lower indices, type (1,2). The rank of a mixed-type tensor is the sum of the number of upper (contravariant) and lower (covariant) indices: 2+1=3 in the above-mentioned example.

The reason for ignoring the contravariant indices in this book lies on the fact that both covariant and contravariant forms are identical in 3D Euclidean cartesian space. Complexities of operations involving contravariant indices are inevitable in fields like the general relativity theorem and non-Euclidean spaces.

6. **The Kronecker delta or the substitution tensor** is the discrete version of the Dirac's delta function. The Kronecker delta is a second-rank symmetric tensor,

$$\delta_{ij} = 0, \quad i \neq j$$
$$\delta_{ij} = 1, \quad i = j \tag{1.40}$$

with the property $\delta_{ij} = \delta_{ji}$.

When the delta function is multiplied by another tensor, one of the indices of the tensor is replaced by one of the indices of the delta tensor. This is the reason we call this symbol the substitution tensor. For example, in $\delta_{ij}u_j = u_i$, the subscript j has been replaced by i. Similarly, in $u_i\delta_{ij} = \delta_{ij}u_i = \delta_{ji}u_i = u_j$, the subscript i is replaced by j. So, the delta tensor helps us to manipulate entries of tensors. It is obvious that $\delta_{ik}\delta_{kj} = \delta_{ij}$. We can prove all these relations using Einstein's summation convention (the next item).

The counterpart of the Kronecker delta in the vector notation is the identity matrix ($\mathbf{I}$). Consequently, translation of $\delta_{ij}u_j = u_i$ to the vector language simply is $\mathbf{Iu} = \mathbf{u}$. The continuous version of this expression can be defined using Dirac's delta function $\int_{-\infty}^{\infty} \delta(x-y)f(x)dx = f(y)$.

7. **The Einstein's summation convention** is a convention we use to simplify mathematical expressions. In this convention, we have to work with repeating (dummy) and non-repeating indices. A symbol that is used to represent a dummy variable can be replaced by any other symbol. Dummy variables like t or x also appear in the integration formula $\int x dx \equiv \int t dt$. However, the selection of non-repeating variables should be consistent in all terms.[7] It should be noted that the number of non-repeating variables of a tensor equals the rank of that tensor. The repeating indices do not change the rank of tensors. For example, τ_{ij} is a second-rank tensor, but τ_{ii} is a scalar.

We have to obey three following rules:

- The summation should only be applied to the dummy variables. Examples are $a_i a_i = \sum_i a_i a_i = a_1 a_1 + a_2 a_2 + a_3 a_3$ or $\delta_{ii} = \delta_{11} + \delta_{22} + \delta_{33} = 1 + 1 + 1 = 3$.[8] Or the matrix product of a 3×3 matrix and a vector $\mathbf{AB} = \sum_j A_{ij}B_j, i = 1, 2, 3$ can simply be written as $A_{ij}B_j$.
- Each dummy index should only appear twice in all terms. For example, the expression $A_{ijk}B_iC_i$ violates this rule, because the index i has appeared three times.
- The non-repeating indices should be identical. For example, the expression $A_{ij}B_j + C_k$ violates this rule because the non-repeating index of the first term (i) and the second term (k) are not identical. In this relation, the index j is a repeating index and the result should be summed over j = 1, 2, 3.

8. **Tensor contraction** is an operation in tensor algebra that results in the reduction of the sum of orders of the inputs by two. In the contraction operation, we set the two neighboring indices of the two tensors equal to each other. This can

be done by multiplying $\delta_{ij} = \delta_{ji}$ to contract indices i and j. Then, based on the summation convention, we have to perform the summation over index i. Contraction converts two non-repeating indices into a repeating index. Since the rank of a tensor should be determined based on the number of non-repeating indices, the order of the result is reduced by 2.

When you want to compute the trace of a matrix, you are performing the contraction on that matrix. The contracted form of a second-order tensor A_{ij} equals the sum of the diagonal components of the matrix, or its trace: $A_{ii} = A_{11} + A_{22} + A_{33}$. The trace of a traceless tensor is zero.

9. **Contracted product of two vectors or the inner product** generates a scalar

$$\mathbf{u} \cdot \mathbf{v} = contraction(u_i v_j) = \delta_{ij} u_i v_j = u_i v_i \qquad (1.41)$$

10. **Contracted product of two matrices**, or simply the matrix product, generates a second-order tensor

$$\mathbf{T} \cdot \mathbf{S} = contraction(T_{ij} S_{kl}) = \delta_{jk} T_{ij} S_{kl} = T_{ij} S_{jl} \qquad (1.42)$$

11. **Double-contracted product** or the double dot product of two tensors is obtained by performing the contraction twice. Hence, the order of the result is reduced by 4. So, the minimum sum of orders of inputs of a double-contraction operator should be 4.[9] For instance, the double dot product of two second-order tensors is a zeroth-order (2+2–4=0) tensor or a scalar,

$$
\begin{aligned}
\mathbf{T} : \mathbf{S} &= contraction(contraction(T_{ij} S_{kl})) = \delta_{jk} \delta_{il} T_{ij} S_{kl} = T_{ij} S_{ji} \\
\mathbf{T} : \mathbf{I} &= T_{ij} \delta_{ij} = T_{ii} \qquad\qquad\qquad\qquad\qquad\qquad (1.43)
\end{aligned}
$$

The second relation denotes that the trace of a matrix equals the double contracted product of the matrix and the identity matrix.

12. **Dot product of a tensor and a vector** is a vector

$$\mathbf{u}.\mathbf{S} = contraction(u_i S_{jk}) = \delta_{ij} u_i S_{jk} = u_i S_{ik} = u_{1S}(1k) + u_{2S}(2k) + u_{3S}(3k) \qquad (1.44)$$

13. **The permutation or the Levi–Civita symbol or the alternating symbol** is a symbol (with three indices) or a third-rank quasi-tensor. Its magnitude equals 1 when the order of its indices is cyclic, equals –1 when the order is acyclic, and otherwise is zero. It has 27 components including 21 zeros, three +1, and three –1 elements. Swapping any two indices reverts the sign of the result.

$$
\begin{aligned}
\varepsilon_{ijk} &= 0, \quad i = j, \ j = k, \ i = k \\
\varepsilon_{ijk} &= 1, \quad Even\ or\ cyclic\ permutation \\
\varepsilon_{ijk} &= -1, \quad Odd\ or\ acyclic\ permutation \qquad (1.45)
\end{aligned}
$$

Even permutations of indices are $(1,2,3)$, $(3,1,2)$, $(2,3,1)$, and odd permutations are $(1,3,2)$, $(3,2,1)$, $(2,1,3)$. Another more compact definition for the permutation symbol is $\varepsilon_{ijk} = \frac{1}{2}(i-j)(j-k)(k-i)$.

The most important application of the permutation symbol is in definitions of the cross product, the curl operator, the determinant, and the dual vector. The permutation symbol generally is not a tensor, since it does not obey the transformation rules between coordinates. The permutation symbol is a tensor just in Euclidean space with an orthonormal basis.

The following three identities including the permutation symbol are very helpful in proving other tensor or vector identities.

a. $\varepsilon_{ijk}\varepsilon_{ijk} = 6$.
b. $\delta_{ij}\varepsilon_{ijk} = \varepsilon_{iik} = 0$.
c. $\varepsilon_{ijk}\varepsilon_{klm} = \delta_{il}\delta_{jm} - \delta_{im}\delta_{jl}$.

Pay attention that the expression $\varepsilon_{ijk}\varepsilon_{ijk}$ is not equal to ε_{ijk}^2, since the former obeys the summation convention while the latter does not. The proof of these identities and also many others will be presented in Section 1.4.

14. **Transpose operation** is obtained when you swap two indices of a tensor,

$$\left(A_{ij}\right)^T = A_{ji} \qquad (1.46)$$

15. **A symmetric tensor** with respect to indices i and j does not change by performing the transpose operation or interchanging indices i and j,

$$\mathbf{A} = \mathbf{A}^T \rightarrow A_{...ij...} = A_{...ji...} \qquad (1.47)$$

An example of a symmetric tensor is the Kronecker delta.[10]

16. **Skew(anti)-symmetric tensor** with respect to indices i and j is defined as

$$\mathbf{A} = -\mathbf{A}^T \rightarrow A_{...ij...} = -A_{...ji...} \qquad (1.48)$$

For a second-rank skew-symmetric tensor, diagonal components are zero and off-diagonal components are opposite in sign: $A_{ij} = 0$ if $i = j$,[11] and $A_{ij} = -A_{ji}$ if $i \neq j$. So, three independent components are enough to define a skew-symmetric matrix. This fact implies that a second-rank skew-symmetric tensor can also be given by a vector. This is similar to the definition of the vorticity vector based on the rotation matrix that will be discussed later. A well-known example of a skew-symmetric tensor is the permutation symbol, which is anti-symmetric with respect to each of its three indices.

Example: Prove that the double-dot product of a symmetric tensor and an anti-symmetric tensor equals 0. Assume that α_{ij} and β_{ij} are anti-symmetric and symmetric second-order tensors, respectively. Consequently, $\alpha_{11} = \alpha_{22} = \alpha_{33} = 0$, $\alpha_{12} = -\alpha_{21}$, $\alpha_{13} = -\alpha_{31}$, $\alpha_{23} = -\alpha_{32}$, $\beta_{12} = \beta_{21}$, $\beta_{13} = \beta_{31}$, $\beta_{23} = \beta_{32}$. In two-dimensional space,

$$\begin{aligned} \alpha_{ij}\beta_{ji} &= \alpha_{11}\beta_{11} + \alpha_{12}\beta_{21} + \alpha_{21}\beta_{12} + \alpha_{22}\beta_{22} \\ &= 0 + \alpha_{12}\beta_{21} - \alpha_{12}\beta_{21} + 0 = 0 \end{aligned} \qquad (1.49)$$

17. **Outer/dyadic product of two vectors** is a second-order tensor defined by

$$(\mathbf{u} \otimes \mathbf{v})_{ij} = u_i v_j \qquad (1.50)$$

18. **Matrix decomposition:** Any second-order tensor can be decomposed into symmetric and anti-symmetric parts

$$W_{ij} = \frac{1}{2}\left(W_{ij} + W_{ji}\right) + \frac{1}{2}\left(W_{ij} - W_{ji}\right)$$

$$\mathbf{W} = \frac{1}{2}\left(\mathbf{W} + \mathbf{W}^T\right) + \frac{1}{2}\left(\mathbf{W} - \mathbf{W}^T\right) \tag{1.51}$$

A similar decomposition can be done for functions or high-order tensors.[12]

19. **Gradient of a scalar** is a vector defined by

$$\nabla\phi = \frac{\partial\phi}{\partial x_j} \tag{1.52}$$

sometimes shown by $\phi_{,j}$.

20. **Gradient of a vector** is a second-order tensor defined by

$$L_{ij} = \nabla\mathbf{u} = \frac{\partial u_i}{\partial x_j} \equiv u_{i,j} \rightarrow \nabla\mathbf{u}^T = \frac{\partial u_j}{\partial x_i} \equiv u_{j,i} \tag{1.53}$$

Example: The velocity gradient as a second-order tensor obeys the matrix decomposition rule. The decomposition of the velocity gradient tensor reads

$$u_{i,j} = \frac{1}{2}\left(u_{i,j} + u_{j,i}\right) + \frac{1}{2}\left(u_{i,j} - u_{j,i}\right) \tag{1.54}$$

The first and the second terms in the right-hand side are the symmetric and the anti-symmetric parts of the velocity gradient, which are called the deformation rate tensor and the rotation tensor, respectively.

21. **Divergence of a vector** is the contracted form of the gradient operator,

$$\nabla\cdot\mathbf{u} = trace(\nabla\mathbf{u}) = contraction\left(\frac{\partial u_i}{\partial x_j}\right) = \frac{\partial u_i}{\partial x_i} = u_{i,i}$$

22. **Divergence of a tensor**

$$\nabla\cdot\mathbf{T} = contraction\left(\frac{\partial T_{ij}}{\partial x_k}\right) = \frac{\partial T_{ij}}{\partial x_i} = T_{ij,i}$$

23. **Cross product of two vectors** is defined using the permutation symbol based on the following relation:

$$(\mathbf{A}\times\mathbf{B})_i = \varepsilon_{ijk}A_jB_k \tag{1.55}$$

24. **Curl of a vector** in three-dimensional space is defined using the permutation symbol

$$\nabla\times\mathbf{A} = \varepsilon_{ijk}\frac{\partial}{\partial x_j}A_k = \varepsilon_{ijk}A_{k,j} \tag{1.56}$$

The structure of the curl operator is similar to the definition of the dual vector.

25. **Curl of a scalar** in three-dimensional space is a second-order tensor

$$\nabla \times \phi = \varepsilon_{ijk} \frac{\partial \phi}{\partial x_j} = \varepsilon_{ijk} \phi_{,j} \tag{1.57}$$

26. **Curl of a second-order tensor** in three-dimensional space is a scalar. For example, the curl of curl of a tensor reads

$$\nabla \times (\nabla \times \boldsymbol{\tau}) = \varepsilon_{ikj} \varepsilon_{lnm} \frac{\partial^2 \tau_{jm}}{\partial x_k \partial x_n} = \varepsilon_{ijk} \varepsilon_{lmn} \frac{\partial^2 \tau_{jm}}{\partial x_k \partial x_n} = \varepsilon_{ijk} \varepsilon_{lmn} \frac{\partial^2 \tau_{jm}}{\partial x_n \partial x_k} \tag{1.58}$$

Two last equalities hold since the order of derivatives of a function could be changed if the function is continuous. The curl of curl of a second-order tensor appears in incompatibility relations in elasticity.

27. **The dual vector of a tensor** is defined using the permutation symbol and is related to the anti-symmetric part of a tensor via the decomposition of the tensor,

$$d_i = \varepsilon_{ijk} T_{kj} = \varepsilon_{ijk} \left(T_{kj}^{sym} + T_{kj}^{anti} \right) = \varepsilon_{ijk} T_{kj}^{anti} \tag{1.59}$$

The last equality holds since the product of the permutation symbol as an anti-symmetric tensor and the symmetric part of the tensor vanishes. So, the dual vector of a tensor is equal to the dual vector of its anti-symmetric part. Since any anti-symmetric tensor includes just three non-zero components, we can construct a vector using these numbers. This vector is the dual vector of the tensor.

Consider the gradient of the velocity vector (**L**),[13] which is a second-rank tensor. On one hand, decompose the velocity gradient into the symmetric and the anti-symmetric parts. Compute the dual vector of **L**. The symmetric part vanishes and the anti-symmetric part (the rotation tensor) survives (**L**^{anti}). On the other hand, you can use three non-zero components of the anti-symmetric part of the velocity gradient tensor in a special manner to make the dual vector of the velocity gradient tensor, which turns out to be the curl of the velocity vector or the vorticity vector (**ω**). As a result, the vorticity vector equals the dual vector of the velocity gradient matrix.

28. **Determinant** is the third invariant of a tensor and is defined as

$$det(\mathbf{A}) = (a_{ij}) = det \begin{pmatrix} a_{11} & a_{12} & a_{13} \\ a_{21} & a_{22} & a_{23} \\ a_{31} & a_{32} & a_{33} \end{pmatrix} = \varepsilon_{ijk} a_{1i} a_{2j} a_{3k} = \frac{1}{6} \varepsilon_{ijk} \varepsilon_{lmn} a_{il} a_{jm} a_{kn} \tag{1.60}$$

Example: Prove that the determinant of a matrix and the determinant of its transpose are equal. Using tensor algebra,

$$det(\mathbf{A}^T) = (a_{ji}) = \varepsilon_{jik} a_{1j} a_{2i} a_{3k} = det(\mathbf{A}) \tag{1.61}$$

Example: Show that swapping two rows of a matrix changes the sign of its determinant. By swapping rows i and k,

$$det(\mathbf{A}_{Swapped}) = \varepsilon_{kji} a_{1i} a_{2j} a_{3k} = -\varepsilon_{ijk} a_{1i} a_{2j} a_{3k} = -det(\mathbf{A}) \tag{1.62}$$

29. **Invariant of vectors:** Invariants are properties of a tensor, which remain unchanged with frame rotation. Vectors only have one invariant, which is equal to the squared magnitude of the vector,

$$I = v_i v_i = \mathbf{v} \cdot \mathbf{v} \tag{1.63}$$

Some vectors like vorticity have a variable length with respect to different frames. Such vectors are called pseudo-vectors. This concept may be extended to tensors to define pseudo-tensors.

30. **Invariants of a tensor** are invariant with respect to the rotation of coordinates. Invariants of a rank-2 tensor ($\mathbf{S}$) are the coefficients of the characteristic polynomial $\lambda^3 - I_1\lambda^2 + I_2\lambda - I_3$ obtained by simplifying the relation $det(\mathbf{S} - \lambda\mathbf{I})$. Eigenvalues ($\lambda$) are the roots of the characteristic polynomial.
The definition of the principal invariants is connected to the eigenvalues of the tensor

$$
\begin{aligned}
I_1 &= trace(\mathbf{S}) = \lambda_1 + \lambda_2 + \lambda_3 \\
I_2 &= \frac{(trace(\mathbf{S}))^2 - trace(\mathbf{S}^2)}{2} = \lambda_1\lambda_2 + \lambda_1\lambda_3 + \lambda_2\lambda_3 \\
I_3 &= det(\mathbf{S}) = \lambda_1\lambda_2\lambda_3
\end{aligned}
\tag{1.64}
$$

There is an infinite number of invariants since we can combine two invariants and make a new one. However, the main invariants of a tensor are[14]

$$
\begin{aligned}
J_1 &= trace(\mathbf{S}) = S_{ii} = I_1 = \lambda_1 + \lambda_2 + \lambda_3 \\
J_2 &= \frac{1}{2}trace(\mathbf{S}^2) = \frac{1}{2}\mathbf{S}:\mathbf{S} = \frac{1}{2}S_{ij}S_{ji} = \frac{I_1^2}{2} - I_2 = \frac{1}{2}\left(\lambda_1^2 + \lambda_2^2 + \lambda_3^2\right) \\
J_3 &= \frac{1}{3}trace(\mathbf{S}^3) = \frac{1}{3}S_{ij}S_{jk}S_{ki} = I_3 - \frac{1}{3}\left(3I_1I_2 - I_1^3\right) = \frac{1}{3}\left(\lambda_1^3 + \lambda_2^3 + \lambda_3^3\right)
\end{aligned}
\tag{1.65}
$$

If the matrix is traceless (like the deviatoric part of the stress tensor, $I_1 = 0$), the first and the second invariants are both zero, $J_2 = -I_2$, $J_3 = I_3$.
The second invariant of tensors is used to define the von-Mises stresses in plasticity, the apparent viscosity for non-Newtonian fluids and powder flows, and to present the Q-criterion in vorticity dynamics based on the second invariant of the velocity gradient tensor.
Example: Find the invariants of the following tensor and check the relations between the invariants and eigenvalues.

$$
\begin{pmatrix}
2 & 0 & 2 \\
0 & 1 & 1 \\
2 & 1 & 3
\end{pmatrix}
\tag{1.66}
$$

First, compute the characteristic polynomial,

$$det \begin{pmatrix} 2-\lambda & 0 & 2 \\ 0 & 1-\lambda & 1 \\ 2 & 1 & 3-\lambda \end{pmatrix} = 0 \rightarrow \lambda^3 - 6\lambda^2 + 6\lambda = 0 \qquad (1.67)$$

Eigenvalues are the roots of this polynomial: 0, $3 + \sqrt{3}$, and $3 - \sqrt{3}$. Based on the coefficients, the first to the third invariants are 6, 6, 0. The first invariant equals the trace of the matrix and also is the sum of eigenvalues. The second invariant equals the sum of product of eigenvalues $0 + 0 + \lambda_2 \lambda_3$. The third invariant equals the determinant of the matrix and is equal to the product of eigenvalues.

31. **Positive definite tensor:** A symmetric[15] tensor (**S**) is positive definite if and only if the expression $\mathbf{S} : \mathbf{v} \otimes \mathbf{v} > 0$ holds for any nonzero vector **v**. It is not easy to determine the positive definiteness of a tensor since we should check all non-zero vectors, which is impossible. A positive definite tensor has three positive eigenvalues or principal values. For example, the pure deformation tensor is positive definite. The concept of positive definiteness appears in the Onsager reciprocal relations in non-equilibrium thermodynamics.

32. **Isotropic tensor** is a non-zero tensor that has the same components with respect to all rotated coordinates. All rank-0 tensors are isotropic. There is no first-rank isotropic tensor. δ_{ij} and ε_{ijk} are the only rank-2 isotropic tensor and rank-3 isotropic symbol, respectively. The most important fourth-order isotropic tensor, which is used in deriving the constitutive relation for momentum transport is

$$A_{ijkm} = B\delta_{ik}\delta_{jm} + C\delta_{im}\delta_{jk} + D\delta_{ij}\delta_{km} \qquad (1.68)$$

Example: Translate the following expressions from vector language to the tensor notation.

- $\mathbf{a} + \mathbf{b} = \mathbf{c}$ equals $a_i + b_i = c_i$.
- $(\mathbf{a} \cdot \mathbf{b})(\mathbf{c} \cdot \mathbf{d}) = 0$ equals $(a_i b_i)(c_j d_j) = 0$.
- $\mathbf{T} = \mathbf{T}'\mathbf{T}''$ equals $T_{ij} = T'_{ik} T''_{kj}$.
- $trace(\mathbf{T}'\mathbf{T}'') = contraction(T'_{ik} T''_{kj}) = T'_{ik} T''_{ki}$
- $\nabla \cdot (\mathbf{u} \otimes \mathbf{u}) = contraction[\frac{\partial}{\partial x_k}(u_i u_j)] = \frac{\partial}{\partial x_j}(u_i u_j)$

A collection of symbols in vector notation, their translation in tensor notation, and their properties have been presented in Table 1.1.

1.4 VECTOR/TENSOR IDENTITIES

In this section, important vector/tensor identities that are frequently used in continuum mechanics are presented and some of them will be proved.

1. Contracted Kronecker-δ in three-dimensional space,

$$\delta_{ii} = \delta_{11} + \delta_{22} + \delta_{33} = 3 \qquad (1.69)$$

In two-dimensions, the result is 2.

Table 1.1

A List of Symbols in Vector Notation and Their Translation in Tensor Notation

Name	Vector Notation	Tensor Notation	Input	Output	Extra Description
Scalar	○	○	-	-	-
Vector	Boldface	O_i	-	-	-
Dyad	Boldface	O_{ij}	-	-	The other name is matrix.
Identity Matrix	I	δ_{ij}	-	-	The tensor form is the Kronecker delta.
Permutation symbol	Determinant of the permutation matrix	ε_{ijk}	-	-	Used to define the cross product, determinant of matrices, the dual-vector, the curl.
Trace	trace()	O_{ii}	Tensor rank-2	Scalar	Contracted form of a 2nd-rank tensor.
Transpose	O^T	$O_{ij} \to O_{ji}$	Rank 2 tensor	Rank 2 tensor	-
Dot product	·	$O_i O_j \to O_{i...} O_{i...}$	Rank at least 2	Decreases rank by 2	Other names: contracted product. The dot product of two vectors is a scalar. The dot product of two rank-2 matrices is a rank-2 matrix (common product of matrices).
Double dot product	:	$O_{ij} O_{lm} \to O_{ij} O_{ij}$	Rank at least 4	Decreases rank by 4	Other names: inner/scalar/double contracted/double divergence product. The double dot product of two rank-2 matrices is the trace of the product of them which is a scalar.
Gradient	∇	$\dfrac{\partial O_{i...}}{\partial x_j}$	Anything	Rank at least 1	Always increases the rank by 1.
Divergence	$\nabla\cdot$	$\dfrac{\partial O_{i...}}{\partial x_i}$	Rank at least 1	Anything	Always decreases the rank by 1. Divergence is the contracted version of the gradient.
Curl	$\nabla\times$	$\varepsilon_{ijk}\dfrac{\partial O_k}{\partial x_j}$	Up to rank n-r-1	3D: $2 \to 0$, $1 \to 1$, $0 \to 2$. 2D: $1 \to 0$, $0 \to 1$. 1D: $0 \to 0$	Is defined based on the Hodge Duality concept. It is not necessarily normal to the input vector. $\nabla\cdot\nabla\times() = 0$.
Laplacian	∇^2 or Δ	$\dfrac{\partial O_{i...}}{\partial x_l \partial x_i}$	Anything	Preserves the rank	Equals $\nabla\cdot\nabla()$. Except in Cartesian coordinates, components of the Laplacian of a vector are not equal to the Laplacian of the components of that vector.
Dyadic product	⊗	$O_i O_j$	Two vectors	Rank 2 tensor	Other names: outer product, tensor product. Sometimes ⊗ is omitted.

2. Product of the Kronecker-δ and the permutation symbol,

$$\delta_{ij}\varepsilon_{ijk} = \varepsilon_{iik} = 0 \tag{1.70}$$

3. The relation between the Kronecker-δ and the permutation symbol,

$$\varepsilon_{ijk} = det \begin{pmatrix} \delta_{i1} & \delta_{i2} & \delta_{i3} \\ \delta_{j1} & \delta_{j2} & \delta_{j3} \\ \delta_{k1} & \delta_{k2} & \delta_{k3} \end{pmatrix} \tag{1.71}$$

Proof: Setting $a \rightarrow \delta, 1 \rightarrow i, 2 \rightarrow j, 3 \rightarrow k$ in Equation (1.60)

$$det \begin{pmatrix} \delta_{i1} & \delta_{i2} & \delta_{i3} \\ \delta_{j1} & \delta_{j2} & \delta_{j3} \\ \delta_{k1} & \delta_{k2} & \delta_{k3} \end{pmatrix} = \varepsilon_{lmn}\delta_{il}\delta_{jm}\delta_{kn} = \varepsilon_{ijk}$$

The last equality holds based on the definition of the substitution tensor.
The other approach to prove this relation is to substitute different sequences of indices on both sides of the relation. We see that if two indices are identical (two similar rows), the determinant will be zero. For even permutation of indices, the determinant reduces to the determinant of the identity matrix that equals 1. It is interesting to show that for the odd permutation of indices, the determinant equals -1. For example

$$\varepsilon_{132} = det \begin{pmatrix} 1 & 0 & 0 \\ 0 & 0 & 1 \\ 0 & 1 & 0 \end{pmatrix} = -1$$

This matrix is called the permutation matrix, which also appears in the numerical solutions of the linear system of equations using the Gaussian elimination method. Hence, the permutation symbol equals the determinant of the permutation matrix. The permutation matrix contains only one 1 in each row and column of the matrix.

4. Product of two permutation symbols

$$\varepsilon_{ijk}\varepsilon_{lmn} = det \begin{pmatrix} \delta_{il} & \delta_{im} & \delta_{in} \\ \delta_{jl} & \delta_{jm} & \delta_{jn} \\ \delta_{kl} & \delta_{km} & \delta_{kn} \end{pmatrix} \tag{1.72}$$

Proof: Using the previous identity and considering the fact that the determinant of the transpose of a matrix equals the determinant of the matrix,

$$\varepsilon_{ijk}\varepsilon_{lmn} = det \begin{pmatrix} \delta_{i1} & \delta_{i2} & \delta_{i3} \\ \delta_{j1} & \delta_{j2} & \delta_{j3} \\ \delta_{k1} & \delta_{k2} & \delta_{k3} \end{pmatrix} \times det \begin{pmatrix} \delta_{l1} & \delta_{l2} & \delta_{l3} \\ \delta_{m1} & \delta_{m2} & \delta_{m3} \\ \delta_{n1} & \delta_{n2} & \delta_{n3} \end{pmatrix}$$

$$= det \begin{pmatrix} \delta_{i1} & \delta_{i2} & \delta_{i3} \\ \delta_{j1} & \delta_{j2} & \delta_{j3} \\ \delta_{k1} & \delta_{k2} & \delta_{k3} \end{pmatrix} \times det \begin{pmatrix} \delta_{l1} & \delta_{l2} & \delta_{l3} \\ \delta_{m1} & \delta_{m2} & \delta_{m3} \\ \delta_{n1} & \delta_{n2} & \delta_{n3} \end{pmatrix}^T$$

The determinant of the product of two square matrices is equal to the product of the determinants,

$$
\varepsilon_{ijk}\varepsilon_{lmn} = det \left[\begin{pmatrix} \delta_{i1} & \delta_{i2} & \delta_{i3} \\ \delta_{j1} & \delta_{j2} & \delta_{j3} \\ \delta_{k1} & \delta_{k2} & \delta_{k3} \end{pmatrix} \times \begin{pmatrix} \delta_{l1} & \delta_{m1} & \delta_{n1} \\ \delta_{l2} & \delta_{m2} & \delta_{n2} \\ \delta_{l3} & \delta_{m3} & \delta_{n3} \end{pmatrix} \right]
$$

$$
= det \begin{pmatrix} \delta_{il} & \delta_{im} & \delta_{in} \\ \delta_{jl} & \delta_{jm} & \delta_{jn} \\ \delta_{kl} & \delta_{km} & \delta_{kn} \end{pmatrix}
$$

The last equality can be obtained after matrix multiplication and using 9 relations similar to the following equality.

$$
\delta_{i1}\delta_{l1} + \delta_{i2}\delta_{l2} + \delta_{i3}\delta_{l3} = \delta_{ik}\delta_{lk} = \delta_{il}
$$

The left-hand side generates 1 if $i = l$ and vanishes if $i \neq l$; so it should equal the Kronecker delta.

5. Contracted product of two permutation symbols. This identity is super helpful to prove many other vector identities.

$$
\varepsilon_{ijk}\varepsilon_{lmk} = \delta_{il}\delta_{jm} - \delta_{im}\delta_{jl} \tag{1.73}
$$

Proof: substitute $n = k$ in the previous identity, use $\delta_{ii} = 3$, and compute the determinant using the first row of the matrix,

$$
\begin{aligned}
\varepsilon_{ijk}\varepsilon_{lmk} &= det \begin{pmatrix} \delta_{il} & \delta_{im} & \delta_{ik} \\ \delta_{jl} & \delta_{jm} & \delta_{jk} \\ \delta_{kl} & \delta_{km} & 3 \end{pmatrix} \\
&= 3\delta_{il}\delta_{jm} - \delta_{il}\delta_{jk}\delta_{km} + \delta_{im}\delta_{jk}\delta_{kl} - 3\delta_{im}\delta_{jl} + \delta_{ik}\delta_{jl}\delta_{km} - \delta_{ik}\delta_{kl}\delta_{jm} \\
&= 3\delta_{il}\delta_{jm} - \delta_{il}\delta_{jm} + \delta_{im}\delta_{jl} - 3\delta_{im}\delta_{jl} + \delta_{im}\delta_{jl} - \delta_{il}\delta_{jm} \\
&= (3 - 1 - 1)\delta_{il}\delta_{jm} - (3 - 1 - 1)\delta_{im}\delta_{jl} \\
&= \delta_{il}\delta_{jm} - \delta_{im}\delta_{jl}
\end{aligned}
$$

6. Double contracted product of two permutation symbols,

$$
\varepsilon_{ijk}\varepsilon_{ljk} = 2\delta_{il}
$$

Proof: substitute $m = j$ in the previous identity.

$$
\begin{aligned}
\varepsilon_{ijk}\varepsilon_{ljk} &= \delta_{il}\delta_{mm} - \delta_{im}\delta_{ml} \\
&= 3\delta_{il} - \delta_{il} = 2\delta_{il}
\end{aligned}
$$

7. Triple contracted product of two permutation symbols,

$$
\varepsilon_{ijk}\varepsilon_{ijk} = 6 \tag{1.74}
$$

Proof: substitute $i = l$ in the previous identity.

$$
\begin{aligned}
\varepsilon_{ijk}\varepsilon_{ijk} = 2\delta_{ii} &= 1 \times 1 + 1 \times 1 + 1 \times 1 + (-1) \\
&\times (-1) + (-1) \times (-1) + (-1) \times (-1) = 6
\end{aligned}
$$

8. The curl field of a vector (like vorticity) is divergenceless,

$$\nabla \cdot (\nabla \times \mathbf{A}) = 0 \tag{1.75}$$

Proof:

$$
\begin{aligned}
LHS \;=\; & \frac{\partial}{\partial x_i}(\nabla \times \mathbf{A})_i = \frac{\partial}{\partial x_i}\left(\varepsilon_{ijk}\frac{\partial}{\partial x_j}A_k\right) = \varepsilon_{ijk}\frac{\partial^2 A_k}{\partial x_i \partial x_j} \\[2mm]
=\; & \frac{\partial}{\partial x_1}\left(\frac{\partial}{\partial x_2}A_3 - \frac{\partial}{\partial x_3}A_2\right) + \frac{\partial}{\partial x_2}\left(\frac{\partial}{\partial x_3}A_1 - \frac{\partial}{\partial x_1}A_3\right) \\[2mm]
+\; & \frac{\partial}{\partial x_3}\left(\frac{\partial}{\partial x_1}A_2 - \frac{\partial}{\partial x_2}A_1\right) \\[2mm]
=\; & 0
\end{aligned}
$$

The term $\varepsilon_{ijk}\dfrac{\partial^2 A_k}{\partial x_i \partial x_j}$ is zero since it equals the double contracted product of an anti-symmetric tensor and a symmetric tensor with respect to i and j.

9. The curl of the cross product of two vectors,

$$\nabla \times (\mathbf{A} \times \mathbf{B}) = \mathbf{A}(\nabla \cdot \mathbf{B}) + (\mathbf{B} \cdot \nabla)\mathbf{A} - (\mathbf{A} \cdot \nabla)\mathbf{B} - \mathbf{B}(\nabla \cdot \mathbf{A}) \tag{1.76}$$

Proof:

$$
\begin{aligned}
[\nabla \times (\mathbf{A} \times \mathbf{B})]_i \;=\; & \varepsilon_{ijk}\frac{\partial}{\partial x_j}(\mathbf{A} \times \mathbf{B})_k = \varepsilon_{ijk}\frac{\partial}{\partial x_j}(\varepsilon_{klm}A_l B_m) \\[2mm]
=\; & \varepsilon_{ijk}\varepsilon_{klm}\frac{\partial}{\partial x_j}(A_l B_m) \\[2mm]
=\; & (\delta_{il}\delta_{jm} - \delta_{im}\delta_{jl})\left(A_l\frac{\partial}{\partial x_j}B_m + B_m\frac{\partial}{\partial x_j}A_l\right) \\[2mm]
=\; & \delta_{il}\delta_{jm}A_l\frac{\partial}{\partial x_j}B_m + \delta_{il}\delta_{jm}B_m\frac{\partial}{\partial x_j}A_l - \delta_{im}\delta_{jl}A_l\frac{\partial}{\partial x_j}B_m \\[2mm]
-\; & \delta_{im}\delta_{jl}B_m\frac{\partial}{\partial x_j}A_l \\[2mm]
=\; & A_i\frac{\partial}{\partial x_j}B_j + B_j\frac{\partial}{\partial x_j}A_i - A_j\frac{\partial}{\partial x_j}B_i - B_i\frac{\partial}{\partial x_j}A_j \\[2mm]
=\; & A_i(\nabla \cdot \mathbf{B}) + (\mathbf{B} \cdot \nabla)A_i - (\mathbf{A} \cdot \nabla)B_i - B_i(\nabla \cdot \mathbf{A})
\end{aligned}
$$

Consider that you may change the indices of the permutation symbol preserving the cyclic or acyclic sequence of indices $\varepsilon_{ijk}\varepsilon_{klm} = \varepsilon_{ijk}\varepsilon_{lmk}$.

10. The gradient of a scalar field like the potential function which is curl-free.

$$\nabla \times \nabla \phi = \mathbf{0} \tag{1.77}$$

11. The gradient of the dot product of two vectors.

$$\nabla(\mathbf{A} \cdot \mathbf{B}) = \mathbf{A} \times (\nabla \times \mathbf{B}) + \mathbf{B} \times (\nabla \times \mathbf{A}) + (\mathbf{A} \cdot \nabla)\mathbf{B} + (\mathbf{B} \cdot \nabla)\mathbf{A} \tag{1.78}$$

12. The cross product of a vector and its curl.

$$\mathbf{A} \times (\nabla \times \mathbf{A}) = \frac{1}{2}(\nabla \mathbf{A})^2 - (\mathbf{A} \cdot \nabla)\mathbf{A} \qquad (1.79)$$

13. The curl of curl of a vector,

$$\nabla \times (\nabla \times \mathbf{A}) = -\nabla^2 \mathbf{A} + \nabla(\nabla \cdot \mathbf{A}) \qquad (1.80)$$

or the vector Laplacian decomposition

$$\nabla^2 \mathbf{A} = \nabla(\nabla \cdot \mathbf{A}) - \nabla \times (\nabla \times \mathbf{A}) \qquad (1.81)$$

14. The curl of the product of a scalar and the gradient of another scalar.

$$\nabla \times (f\nabla g) = \nabla f \times \nabla g + f\underbrace{\nabla \times \nabla g}_{\mathbf{0}} = \nabla f \times \nabla g \qquad (1.82)$$

15. The divergence of the product of a scalar and a vector.

$$\nabla \cdot (\phi \mathbf{u}) = \phi \nabla \cdot \mathbf{u} + \mathbf{u} \cdot \nabla \phi \qquad (1.83)$$

16. The curl of Laplacian of a vector.

$$\nabla \times (\nabla^2 \mathbf{u}) = \nabla^2 (\nabla \times \mathbf{u}) \qquad (1.84)$$

1.5 INTEGRAL THEOREMS

A list of common theorems in continuum mechanics and related fields involved with integrals of continuous functions is presented here.

1. **The Stokes–Cartan or the General Stokes theorem** is the general form of all integral theorems expressed for manifolds and is used in topology. The deep discussion of this theorem is out of the scope of the present book.

2. **The Curl (the Kelvin-Stokes) theorem** is used in the definition of circulation based on vorticity, and the definition of conservative forces. The curl theorem converts the surface integral of the curl of a quantity to the line integral of that quantity along the closed boundaries of the surface.

$$\int_S (\nabla \times \mathbf{u}) \cdot \hat{\mathbf{n}} dS = \oint_c \mathbf{u} \cdot \hat{\mathbf{t}} dc \qquad (1.85)$$

In addition, the direction of the line integration is determined by the right-hand rule, which indicates that when the thumb is parallel to the direction of the normal vector, the positive orientation of c is the direction of the curling of fingers.

3. **Green's theorem:** In cartesian Euclidean three-dimensional space, for a continuous vector $\mathbf{u} = (P, Q, R)$, the Curl theorem simplifies to

$$\oint_c (Pdx + Qdy + Rdz) =$$
$$\int_S \left[\left(\frac{\partial R}{\partial y} - \frac{\partial Q}{\partial z} \right) dydz + \left(\frac{\partial P}{\partial z} - \frac{\partial R}{\partial x} \right) dzdx + \left(\frac{\partial Q}{\partial x} - \frac{\partial P}{\partial y} \right) dxdy \right] \tag{1.86}$$

Green's theorem is the two-dimensional version, which relates the double integral over a plane region to the line-integral along the curve surrounding the surface.

$$\oint_c (Pdx + Qdy) = \int_S \left(\frac{\partial Q}{\partial x} - \frac{\partial P}{\partial y} \right) dxdy \tag{1.87}$$

4. **The divergence (Gauss's) theorem** is used in the derivation of the weakly compressible form of the Navier–Stokes equation and in obtaining the integral form of the conservation laws from their differential form (the Reynolds' transport theorem). The divergence theorem relates the volume integral of the divergence of a quantity to the surface integral of the flux of that quantity through the closed surface bounding the volume.

$$\int_\forall \nabla \cdot \mathbf{u} d\forall = \oint_S \hat{\mathbf{n}} \cdot \mathbf{u} dS \tag{1.88}$$

The divergence of a vector is a scalar. Hence this equation is a single scalar equation.

5. **The gradient theorem** converts the volume integral of the gradient of a quantity into the surface integral of the product of that quantity and the normal outward unit vector of the surface bounded the volume,

$$\int_\forall \nabla\phi d\forall = \oint_S \phi\hat{\mathbf{n}} dS \tag{1.89}$$

where ϕ is a scalar and $\hat{\mathbf{n}}$ is the normal outward vector of the surface S which bounds the volume $\forall$. Consider that the gradient of a scalar is a vector. So, the first equation is a vector equation consisting of 3 components.

6. **The Laplace theorem (Green's first identity)** is used in the derivation of the weak form of the Navier–Stokes equations. The Green's first identity is the high-dimensional version of the integration by part,

$$\int_S (\psi\nabla^2\phi + \nabla\psi \cdot \nabla\phi)dS = \oint_c \psi(\nabla\phi \cdot \hat{\mathbf{n}})dc \tag{1.90}$$

where c is the boundary of S, $\hat{\mathbf{n}}$ is the outward unit normal vector of the surface element dS.

7. **Leibniz's formula** is used in the derivation of the integral form of the conservation laws based on Reynolds' transport theorem. Leibniz's formula helps us

interchange the order of integration and differentiation,

$$\frac{d}{dt}\int_{\forall(t)}\phi d\forall = \int_{\forall(t)}\frac{\partial\phi}{\partial t}d\forall + \int_{S(t)}\phi\mathbf{u}_s\cdot\hat{\mathbf{n}}dS \tag{1.91}$$

where ϕ is a general tensor field and $\mathbf{u}_s$ is the velocity of change of $S(t)$. It can be concluded that if the integrand is a continuous function and the boundary of integration is not deformable ($\mathbf{u}_s = \mathbf{0}$) or the boundary velocity vector is normal to the boundary ($\mathbf{u}_s\cdot\hat{\mathbf{n}} = 0$), we can interchange the order of integration and differentiation.

8. **The fundamental theorem of calculus** relates the integration of a function (f) to its primitive function (F),

$$\int_a^b f(x)dx = F(b) - F(a)$$

$$\frac{dF(x)}{dx} = f(x) \tag{1.92}$$

9. **The Schwarz' theorem** speaks about the condition under which we can interchange the order of differentiations (symmetry of the second mixed derivatives). Although this theorem does not contain integral, it is a very important theorem in continuum mechanics. If f is a scalar with continuous second-order derivatives, then $\frac{\partial^2 f}{\partial x_i \partial x_j}$ is a symmetric second-order tensor:

$$\frac{\partial^2 f}{\partial x_i \partial x_j} = \frac{\partial^2 f}{\partial x_j \partial x_i} \tag{1.93}$$

1.6 EXERCISES

1.1 Prove that in three-dimensional space, the double-dot product of a symmetric tensor and an anti-symmetric tensor equals 0.

1.2 Obtain the following relation for the gradient of the dot product of two vectors.

$$\nabla(\mathbf{A}\cdot\mathbf{B}) = \mathbf{A}\times(\nabla\times\mathbf{B}) + \mathbf{B}\times(\nabla\times\mathbf{A}) + (\mathbf{A}\cdot\nabla)\mathbf{B} + (\mathbf{B}\cdot\nabla)\mathbf{A} \tag{P1.1}$$

1.3 Obtain the following relation for the curl of the product of a scalar and the gradient of another scalar.

$$\nabla\times(f\nabla g) = \nabla f\times\nabla g \tag{P1.2}$$

1.4 If $\mathbf{F}$ is a second-order anti-symmetric tensor, and $\mathbf{M}$ is a vector equal to $0.5\varepsilon_{ijk}F_{jk}$, find the product $\varepsilon_{pqi}M_i$.

1.5 Write down the velocity gradient tensor and obtain its dual vector. Repeat this procedure to obtain the dual vector of the anti-symmetric part of the velocity gradient tensor. Show that both results are identical.

1.6 Use the conclusion of the previous exercise to prove

$$\boldsymbol{\omega} \times \mathbf{u} = 2\mathbf{L}^{anti} \cdot \mathbf{u},$$

where $\mathbf{u}$ is velocity (a vector), $\boldsymbol{\omega}$ is the curl of $\mathbf{u}$ (vorticity), and $\boldsymbol{\omega} \times \mathbf{u}$ is the Lamb vector.

1.7 If $\mathbf{A}$ is a second-order constant tensor and $\mathbf{x}$ is a vector in three-dimensional space, compute the result of $\nabla \cdot (\mathbf{A}\mathbf{x}\mathbf{x})$.

1.8 If $\mathbf{A}$ is a second-order symmetric constant tensor and $\mathbf{x}$ is a vector, compute the result of $(A_{ij}x_ix_j)_{,k}$.

1.9 Compute the determinant of dyadic product of two vectors $|\mathbf{a} \otimes \mathbf{b}|$.

1.10 Compute $\oint_S x_i n_j dS$ where n_j is the normal unit vector of S which bounds the volume $\forall$.

1.11 Compute $\frac{\partial A_{ij}}{\partial x_k} \delta_{ij}$ if $A_{ij} = x_i x_j$.

1.12 If $S_{ij} + 3S_{kk}\delta_{im}\delta_{jm} = T_{ij}$, then obtain a relation for S_{ij}.

1.13 If A_{ij} is an anti-symmetric constant tensor and $c_i = \varepsilon_{ijk}A_{jk}$ is a vector, then compute

$$\int \varepsilon_{pqi}c_i d\forall$$

NOTES

1. Important exceptions are shells and two-dimensional interfaces between two immiscible fluids along which the density jumps and the two-dimensional space is non-Euclidian.

2. This property of the gradient operator is used in gradient-based continuous neural network methods to find local minimum points.

3. The Hessian of a scalar function is defined as the Jacobian of its gradient.

4. An example of this case is vorticity in two-dimensional flow in the xy plane, which can be described by only one scalar, ω_z.

5. There are several letters exchanged between Einstein and Levi-Civita. We only have access to the responses written by Einstein. The author strongly recommends reading such historical documents.

6. Einstein indicated that the physical governing equations obtained from the conservation laws are independent of coordinate systems. This property of tensors helps us truly derive such equations.

7. For example, Zahra is a common name among Iranians. It is rather expected that you have two or even three friends with the name Zahra. So, you need nicknames to prevent getting confused when you call them. Dummy variables are nicknames of mathematical symbols.

8. This is the first tensor identity you have just proved!

9. Sometimes the double dot product is defined as $\tau_{ij}\tau'_{ij}$, which equals $\boldsymbol{\tau} : \boldsymbol{\tau}'^T$ based on our definition.

10. For higher-order tensors, minor and major symmetries can be defined: $T_{ijkl} = T_{jikl}$ and $T_{ijkl} = T_{klij}$.

11. $A_{ij} = 0$ if $i = j$ should not be written as A_{ii}, because it will wrongly obey the summation convention.

12. For a third-order tensor the anti-symmetric part with respect to all indices is
$W_{ijk}^{anti} = \frac{1}{3!}\left(W_{ijk} - W_{ikj} + W_{jki} - W_{jik} + W_{kij} - W_{kji}\right)$

13. In some references the velocity gradient tensor is defined as $L_{ij} = \partial_{(i}u_{j)}$. Following this convention, the definition for the dual vector should be replaced by $d_i = \varepsilon_{ijk}T_{jk}$. Also, in some textbooks, the dual vector is defined as $d_i = \frac{1}{2}\varepsilon_{ijk}T_{kj}$.

14. Similar to the second invariant, the Frobenius norm of a tensor shown by $||.||_F$ is defined as

$$||\mathbf{S}||_F = \sqrt{\mathbf{S}^T : \mathbf{S}} = \sqrt{trace(\mathbf{S}^T\mathbf{S})} = \sqrt{S_{ij}S_{ij}}$$

and the 2-norm of $\mathbf{S}$ equals the square root of the largest absolute value of eigenvalues (spectral radius) of $\mathbf{S}^T\mathbf{S}$.

15. This definition can be extended to non-symmetric matrices.

2 Preliminary Concepts

Continuum mechanics is a branch of science that deals with the derivation of governing equations in different diverse applications under the umbrella of continuum principle. Such equations are expressed using the tensor/vector notation language. In continuum mechanics, we pay attention to the continuum behavior of material rather than its discrete structure of particles. The relation of continuum mechanics with other branches of science from the viewpoint of solution methods and physical results is shown in Figure 2.1. As is seen in the figure, continuum mechanics is the heart of classical theoretical physics.

The history of the evolution of mechanical sciences consists of two branches: experimental and theoretical developments. The theoretical branch has two sub-branches of atomic-based (microscopic nature)[1] and the continuum-based (macroscopic nature) modelings. Statistical physics employs the probability theory to relate macroscopic and microscopic events (mesoscopic nature).

Continuum mechanics, first formulated by Augustin–Lois Cauchy during the 19th century, tries to ignore atoms. Continuum mechanics supposes that material continuously fills the space. For instance, suppose that we are going to model the motion of cars in the streets of a city. From a molecular non-continuum viewpoint, you may follow the trajectory of each separate car. You try to find a way to compute the interaction of drivers with each other and anticipate their decisions. But, in continuum modeling, we present a partial differential equation containing some self-defined properties, which can be solved to simulate the traffic as a continuous flow of cars. In continuum modeling, we suppose that cars have occupied all spaces in streets which is somehow true in heavy traffics.

2.1 WHAT IS CONTINUUM MECHANICS?

1. **Continuum hypothesis:** Continuum mechanics has been established based on the continuum hypothesis. Simply speaking, the continuum hypothesis states that we can ignore atoms when the length-scale of the problem (L) is sufficiently greater than the inter-atomic spacing or the mean-free-path (λ). To quantify this point, we define the dimensionless Knudsen number

$$Kn = \frac{\lambda}{L} \tag{2.1}$$

When the Knudsen number increases, the continuum hypothesis gradually is violated. So, if the mean-free-path increases like what happens in rarified gas dynamics, or the characteristic length of the problem decreases in nanoscale gaseous flows, the continuum hypothesis becomes invalid.

The validity of the continuum principle paves the road for defining various material properties such as density, mass, and temperature as continuous

DOI: 10.1201/9781032719405-2

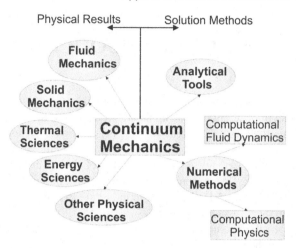

Figure 2.1 Continuum mechanics as the heart of physical and mathematical sciences.

mathematical functions. It means that the definition of a property such as pressure for a non-continuum medium is meaningless. The matter has no sensation about temperature, stress, or density. These properties help us to formulate the governing equations in a continuum framework. From the mathematical viewpoint, different quantities in continuum mechanics are continuous functions, and hence, we are allowed to use differential calculus to derive governing equations from conservation laws containing integrals or derivatives.

2. **Ergodicity and homogeneity:** The continuum principle accurately builds a connection between experimental and theoretical studies when microstructures of the system under consideration are statistically homogenous and ergodic.

The homogeneity of a system speaks about the statistical properties of each part of an overall system that should be the same. A process is ergodic if the statistical properties of the system can be obtained from a random sample of the process. If the process is not ergodic, it means that the statistical behavior of the process changes dramatically even in long-enough samples. The Brownian motion of particles and flipping a coin are well-known examples of an ergodic process, and some financial trends may not be ergodic.

In other words, the ergodic hypothesis states that the behavior of the system is constant when it is averaged over time and all states of the system in its phase space. This hypothesis was presented by the efforts of Ludwig Boltzmann, Henri Poincare, and Willard Gibbs. The phase space is created by the collection of all possible subcases from a bird's eye view. For instance, when you are flipping a coin the phase space contains all possible results.

As an example, consider a series of letters a, e, i, o, h, w, z. If you try to guess the next letter in this sequence, you need to find the hidden connection between these data, if exists. Maybe, these letters are a sequence of vowels, or the starting letters of the names of my favorite foods, or anything else. What if there is a mistake during the report of these data, or maybe they are written randomly, without any particular trend. If this is true, it means that the process is non-ergodic. So, you cannot focus on a small portion of the system to find a general trend to predict its behavior.

3. **The representative elementary volume:** Ergodicity and homogeneity of material allow us to investigate the behavior of a small part of the body as a representative of the overall medium. So, all measurements and computations are made on the representative elementary volume (REV), and the result will be extended to the whole body. So, REV is a small enough part of the matter for which the macroscopic continuum representation is accurate enough to deal with the averaged responses.

The material properties are generally a function of the size or the Knudsen number. The size-dependent properties are called the film properties that are different from bulk properties in non-continuum scales. It means that in continuum mechanics, the material properties and the governing equations are not functions of size of the body and the configuration of its boundaries. A continuum is a body that can be consistently divided into small infinitesimal sub-elements without experiencing any change in properties in comparison to those of the bulk material.

The Hill–Mandel condition indicates that the representative elementary volume is statistically an indicator of behavior of an entire ergodic and homogenous medium. It should carry an adequate number of molecules so that the constitutive relations become independent of the boundary conditions. Consequently, the Hill–Mandel condition allows us to define the constitutive relations based on macroscopically defined properties.

4. **Conservation and balance laws** are fundamental laws of physics, which speak about the rate of change of some measurable properties. Based on Noether's theorem, each symmetry of a system has an associated conservation law; like symmetry with respect to time, space, and rotation, which lead to the conservation of energy, linear momentum, and angular momentum, respectively. We divide the conservation laws into absolute (or exact) and partial (or approximate) laws. The exact laws always hold, but, partial laws may be violated under some particular circumstances.

The conservation of mass or conservation of energy in cosmic scales are partial laws. The mass–energy equivalence defined in the special relativity theorem states that inertia is a function of the content of energy. The mass in the conservation of mass is sometimes called the rest mass. The rest mass is the mass of a body when its momentum is zero. The relativistic mass or the total mass is defined when the body is moving with a velocity equal to a portion of the light velocity. Examples of other common conservation laws mostly appearing in thermo-fluid mechanics are a balance of entropy/exergy[2] and conservation of circulation, vorticity, vehicles, and salinity.

Conservation laws may be written in weak (or global) integral form or strong (local) differential form. A global conservation law holds when we are dealing with an overall amount of a quantity integrated over the whole universe or a finite volume. But, the local laws require the conservation of a quantity at a point in space or for an infinitesimal element when its volume goes to zero. The local laws are called strong since when a quantity is conserved at a local point, it will also be conserved all over a finite volume. Moreover, four laws of thermodynamics and three of Newton's laws could be added to the list of governing equations.

5. **Constitutive relations:** It is mentioned that the main fundamental equations of science are conservation laws. However, the conservation laws are not enough to close governing equations. Some extra information about the behavior of the material is needed through constitutive laws. Such relations are phenomeno-logical equations and may be violated. Then, another question arises here: how can we be sure about the accuracy of the results obtained from the solution of equations derived using the constitutive relations? The answer to this equation lies on the second law of thermodynamics. The second law is responsible to check the validity of constitutive relations in different fields of continuum mechanics. So, each constitutive relation should pass the second law test.

In other words, the constitutive relations help us to make the conservation laws meaningful and applicable. For example, the conservations of momentum and energy speak about the balance of stress and heat flux over a differential element. But, heat flux and stress do not directly make sense in real applications. Instead, we are likely to work with velocity and temperature, which are more familiar and measurable quantities.

As a result, the constitutive relations try to act as a bridge between the conservation laws and the material properties, under the umbrella of the second-law test. Researchers consistently try to find some corrected or modified governing equations in the context of continuum mechanics to predict the non-continuum behaviors that are not predictable by the traditional equations of continuum mechanics.

Different constitutive laws are the source of various thermo–electro–hydro–mechanical behavior of the material. Although the main governing equation of thermal transport in living tissues, porous media, or nanoscale geometries is the conservation of energy, the distribution of temperature and heat flux differ in these cases. Then, what is the difference between equations of motion for powder, air, glue, honey, or yogurt? All these flow configurations obey the same conservation of momentum rule, but the origin of differences in their behavior is their non-identical constitutive relations.

Some well-known phenomenological relations in diverse fields of science are listed here.

- The Stokes' relation between the stress tensor and the strain-rate tensor in fluid mechanics or powder flows.
- The Hook's relation in elasticity between the stress tensor and the strain tensor.
- The Fick's law of diffusion for mass transfer.
- The Fourier's law for heat transfer.

- The Darcy's law for flow in porous media.
- Constitutive relations for conservation of angular momentum.
- Constitutive relations for gradient materials.
- Kinetic equation of state as a relation between pressure, density, temperature, and equations such as $e = c_v T$ as the caloric equation of state to compute internal energy as a function of temperature.
- The Laplace's relation for the surface tension-involving flows.
- Phenomenological equations like the Onsager reciprocal relations and the Curie principle. In equilibrium thermodynamics, the thermodynamic forces and the resulting fluxes are both zero. Out of equilibrium condition, we may assume some linear phenomenological relations or extend them to construct nonlinear relations between fluxes and forces.
- Constitutive relations in electrodynamics are Ohm's law between the electric current and the electric field, the relation between the electric displacement and the electric field in dielectrics, and the relation between the magnetic induction and the magnetic field.
- The relation between the temperature gradient and the current density in thermoelectric effect, including the Peltier and the Seebeck effects.
- The Stefan–Boltzmann law for black bodies in radiative heat transfer.
- Constitutive laws in photonics including the Planck's law and the definition of the speed of light.

2.2 FUNDAMENTAL CONCEPTS

In elementary solid mechanics, we define stress as the ratio of force and area. But, how can we divide two vectors? But, we know how to multiply a second-order tensor ($\boldsymbol{\tau}^c$) by a vector ($\mathbf{A}$), which results the force vector ($\mathbf{F}$),

$$\boldsymbol{\tau}^c = \frac{\mathbf{F}}{\mathbf{A}} \rightarrow \mathbf{F} = \boldsymbol{\tau}^c \cdot \mathbf{A} \tag{2.2}$$

Then, other questions arise: what is the physical interpretation behind this definition? Why is the stress a second-order tensor? In this section, we try to answer such questions by introducing fundamental concepts of continuum mechanics.

1. **Matrices as transformations:** The stress tensor is among the most fundamental elements in continuum mechanics. We need to look at matrices as transformations based on the dot product of a matrix and a vector. When a matrix as a machine is multiplied by a vector (input), it generates another vector as the output of the machine. For example, the stress tensor produces the traction vector at a point on a plane when the normal unit vector of the plane is the input of the machine. Or the Jacobian matrix helps us to transform the material coordinate to the Eulerian frame. The determinant of a transformation matrix is the ratio of the transformed volume and the initial volume.

2. **Galilean invariance:** Galileo Galilei stated that the governing equations should remain unchanged with respect to the coordinates moving with constant speeds.[3] As a result, the governing equations of splashing water in a glass on your table in a train should remain the same when the train is stationary or moves at a constant speed. The material derivative operator is also Galilean invariant.

3. **The Euler–Cauchy stress principle:** Suppose that you are dividing an imaginary volume into two parts by cutting along a surface. Two new opposite internal surfaces appear.[4] Depending on the direction of motion of the knife and based on the right-hand rule, these two surfaces have opposite normal vectors.

4. **The traction vectors:** The Euler–Cauchy stress principle states that the stress on a surface is a vector and depends on the unit normal vector of the surface. So, we tag the stress vector or the traction vector of a point on a surface with the superscript "n", referring to dependence on the normal direction of the surface. The stress state of the matter can be defined based on traction vectors. Projections of the stress vector $\mathbf{T}^{(n)}$ along the normal and the tangential unit vectors are scalars called the normal (τ_n) and the shearing stresses (τ_{sh}), respectively,

$$\tau_n = \mathbf{T}^{(n)} \cdot \mathbf{n}$$
$$\tau_{sh} = \sqrt{\left(\mathbf{T}^{(n)}\right)^2 - \tau_n^2} \qquad (2.3)$$

5. **Cauchy's postulate**[5] indicates that the stress vector at a specific point, $\mathbf{T}^{(n)}$, is the same for all surfaces with identical normal vectors passing through that point. Cauchy's postulate states that the traction vector is a function of the normal vector of the plane regardless of the curvature of the surface.

6. **Cauchy's reciprocal theorem** states that the stress vectors acting on a pair of cutting surfaces are equal in magnitude, but opposite in direction. So, they can be treated as an action-reaction couple, based on Newton's third law

$$\mathbf{T}^{(-n)} = -\mathbf{T}^{(n)} \qquad (2.4)$$

7. **The Cauchy's stress theorem:** In accordance with the Euler–Cauchy stress principle, the state of stress at a point depends on normal vectors of an infinite number of cutting planes passing through the point. Cauchy's stress theorem expresses that if you know the stress vector on three perpendicular planes among infinite number of planes, you can compute the stress vector on any other plane passing through that point.

8. **The Cauchy's stress tensor ($\boldsymbol{\tau}^c$):** The traction vector at a point with the normal unit vector of $\mathbf{n}$ is a function of the position of the point and time. This means that the traction (stress) field at a point in a continuum is a function of time, position, and orientation of the internal cutting surfaces (the normal vector of the plane), $\mathbf{T}^{(n)} = f(t, \mathbf{x}, \mathbf{n})$. It is obvious that the traction field is a function of two vectors, and hence, the stress is not expected to be a vector.

Based on the Cauchy's stress theorem, we need the stress vector on three orthogonal planes to completely determine the state of stress. So, if nine (3×3) stress components along three normal directions are given, we may obtain the traction vector in any other direction. Consequently, we can define the stress tensor with nine elements $\boldsymbol{\tau}^c(\mathbf{x},t)$ that is not a function of the normal vector anymore. Therefore, the traction vector in a particular direction can be derived using

$$\mathbf{T}^{(n)} = \boldsymbol{\tau}^c \cdot \mathbf{n} \quad or \quad T_j^{(n)} = \tau_{ij}^c n_i \tag{2.5}$$

Equation (2.5) implies that we can easily obtain the stress vector on a surface by computing the dot product of the normal vector of the plane and the stress tensor. If the traction vector on a surface is zero, that surface is called traction-free. In other words, you may compute the traction vector on an arbitrary plane with the normal vector (n_1, n_2, n_3) by a linear combination of traction vectors on three perpendicular planes,

$$\mathbf{T}^{(n)} = n_1 \begin{pmatrix} \tau_{11}^c \\ \tau_{21}^c \\ \tau_{31}^c \end{pmatrix} + n_2 \begin{pmatrix} \tau_{12}^c \\ \tau_{22}^c \\ \tau_{32}^c \end{pmatrix} + n_3 \begin{pmatrix} \tau_{13}^c \\ \tau_{23}^c \\ \tau_{33}^c \end{pmatrix}$$

$$= \begin{pmatrix} \tau_{11}^c & \tau_{12}^c & \tau_{13}^c \\ \tau_{21}^c & \tau_{22}^c & \tau_{23}^c \\ \tau_{31}^c & \tau_{32}^c & \tau_{33}^c \end{pmatrix} \begin{pmatrix} n_1 \\ n_2 \\ n_3 \end{pmatrix} = \boldsymbol{\tau}^c \cdot \mathbf{n} \tag{2.6}$$

To find the normal stress, we need to compute the dot product of the traction vector and the normal unit vector: $\tau_n = (\boldsymbol{\tau}^c \cdot \mathbf{n}) \cdot \mathbf{n}$. This relation is the reason why the diagonal components of the stress tensor should indicate the normal stresses.

9. **Cauchy's first law:** The main duty of the stress tensor is to transmit the stress state on boundaries into the volume of the domain using the traction vectors and mechanical connections between the mentioned cutting surfaces. Then, the surface forces obtained from the stress tensor will be coupled with the volume(body) forces within the framework of the conservation laws.

   Based on the conservation of momentum or the second law of Newton for a continuum body without acceleration, Cauchy's stress tensor should satisfy the following differential equation:

$$\frac{\partial \tau_{ji}^c}{\partial x_j} + F_i = 0 \tag{2.7}$$

   where the first term on the left-hand side is the divergence of the stress tensor (the surface force density) and F_i denotes the body force per unit volume.

   For the hydrostatic limiting case, after decomposing the stress tensor into the deviatoric and the hydrostatic parts, it is straightforward to show that the stress field is $\tau_{ij}^c = -p\delta_{ij}$.

10. **The Navier's equation of motion** is easily obtained if we include the acceleration (the second-order time-derivative of the displacement vector, d_i) to the

right-hand side of the Cauchy's first law,

$$\frac{\partial \tau_{ji}^c}{\partial x_j} + F_i = \rho \frac{\partial^2 d_i}{\partial t^2} \tag{2.8}$$

It can be concluded that if there is no acceleration and no body forces, the stress field is divergenceless.

11. **The Cauchy's second law** speaks about the simplified form of the conservation of angular momentum at a point. This law indicates that the stress tensor is symmetric, and three off-diagonal pairs of shear stresses are equal. So, we are facing only 3 normal and 3 shear components in Cauchy's stress tensor.

$$\tau_{ij}^c = \tau_{ji}^c \tag{2.9}$$

In applications such as polar materials and some special types of non-Newtonian fluids or when an external body torque exists, the stress tensor is asymmetric.

12. **Decompositions:** There are three common decompositions in continuum mechanics:

    a. Decomposition of the velocity gradient tensor into the symmetric part (the deformation-rate tensor) and the anti-symmetric part (the spin or the rotation tensor) as discussed after Equation (1.54).

    b. Polar decomposition of the deformation gradient tensor with non-zero determinant ($\mathbf{F}$), into the product of two tensors $\mathbf{F} = \mathbf{R} \cdot \mathbf{U}$ or $\mathbf{F} = \mathbf{L} \cdot \mathbf{R}$, in which $\mathbf{R}$ is the orthogonal rotation tensor with the determinant equal to 1 (without volume change), $\mathbf{U}$ and $\mathbf{L}$ are symmetric positive-definite right and left stretching tensors, respectively.

    This decomposition indicates that a deformation is a combination of two successive transformations: a stretch then a rotation or a rotation then a stretch. $\mathbf{C} = \mathbf{F}^T \mathbf{F}$ and $\mathbf{B} = \mathbf{F}\mathbf{F}^T$ are the right and the left Cauchy–Green deformation tensors, respectively.

    c. Deviatoric decomposition: The Cauchy's stress tensor τ_{ij}^c can be written as the sum of the hydrostatic (volumetric/mean normal/spherical) stress tensor $\pi\delta_{ij}$, and the deviatoric (viscous) stress tensor τ_{ij}.

$$\begin{aligned} \tau_{ij}^c &= \tau_{ij} + \pi\delta_{ij} \\ \pi &= \frac{\tau_{kk}^c}{3} = \frac{\tau_{11} + \tau_{22} + \tau_{33}}{3} = \frac{1}{3}I_1 \end{aligned} \tag{2.10}$$

where I_1 is the first invariant of the stress tensor. So,

$$\begin{pmatrix} \tau_{xx}^c & \tau_{xy}^c & \tau_{xz}^c \\ \tau_{yx}^c & \tau_{yy}^c & \tau_{yz}^c \\ \tau_{zx}^c & \tau_{zy}^c & \tau_{zz}^c \end{pmatrix} = \begin{pmatrix} \tau_{xx} & \tau_{xy} & \tau_{xz} \\ \tau_{yx} & \tau_{yy} & \tau_{yz} \\ \tau_{zx} & \tau_{zy} & \tau_{zz} \end{pmatrix} + \begin{pmatrix} \pi & 0 & 0 \\ 0 & \pi & 0 \\ 0 & 0 & \pi \end{pmatrix} \tag{2.11}$$

If the hydrostatic part or the deviatoric part vanishes, we call the stress tensor purely deviatoric or purely spherical, respectively. This decomposition can be used to extract a traceless tensor from another tensor, like what we do to produce a traceless sub-grid-scale tensor in large-eddy simulation.

The first invariant of the deviatoric part of the stress tensor, namely D_{ij}, is zero. The second invariant equals $I_{2D} = -\frac{1}{3}I_1^2 + I_2$, and the third invariant is $I_{3D} = I_3 - \frac{1}{3}I_1 I_2 + \frac{2}{27}I_1^3$. The equivalent stress equals $\sqrt{-3I_{2D}}$ where I_{2D} is the second invariant of the deviatoric part. The first invariant of the spherical part equals the first invariant of the stress tensor, the second invariant of the spherical part is $\frac{1}{3}I_1^2$, and the third invariant of the spherical part is $\frac{1}{27}I_1^3$.

The octahedral stresses are normal and shear stresses on a plane with its normal vector making equal angles with the principal directions when the stress tensor is diagonal. In other words, the octahedral stresses are obtained based on the normal vector of one-eighth of an octahedron with the center located at the origin. The normal unit vector of such surface is $(\frac{1}{\sqrt{3}}, \frac{1}{\sqrt{3}}, \frac{1}{\sqrt{3}})$. Consider the stress tensor in the form

$$\begin{pmatrix} \tau_{11}^c & 0 & 0 \\ 0 & \tau_{22}^c & 0 \\ 0 & 0 & \tau_{33}^c \end{pmatrix}$$

Consequently, the octahedral normal stress is $\tau_n^{oc} = (\boldsymbol{\tau}^c \cdot \mathbf{n}) \cdot \mathbf{n} = \frac{\tau_{ii}^c}{3} = \frac{I_1}{3}$ or the hydrostatic stress. The octahedral shear stress is related to the second invariant of the deviatoric stress tensor, which appears in non-Newtonian mechanics and the theory of plasticity. The octahedral shear stress equals $\tau_{sh}^{oc} = \sqrt{(\boldsymbol{\tau}^c \cdot \mathbf{n})^2 - (\tau_n^{oc})^2} = \sqrt{-\frac{2}{3}I_{2D}} = \frac{1}{3}\sqrt{2I_1^2 - 6I_2}$.

Example: Compute the spherical and the deviatoric parts of the following tensor and their invariants. Check the relations between the invariants of the deviatoric and the spherical parts and the invariants of the tensor.

$$\begin{pmatrix} 2 & 0 & 2 \\ 0 & 1 & 1 \\ 2 & 1 & 3 \end{pmatrix}$$

The first, the second, and the third invariants of this tensor are 6, 6, and 0, respectively. Compute the octahedral normal and shear stresses and the equivalent stress.

The average of the diagonal components is $\pi = 2$. So the deviatoric and the spherical parts are

$$\mathbf{D} = \begin{pmatrix} 0 & 0 & 2 \\ 0 & -1 & 1 \\ 2 & 1 & 1 \end{pmatrix} \quad \mathbf{S} = \begin{pmatrix} 2 & 0 & 0 \\ 0 & 2 & 0 \\ 0 & 0 & 2 \end{pmatrix}$$

The characteristic polynomial of the deviatoric part is $\lambda^3 - 6\lambda - 4$ and the eigenvalues are $(-2, 1 + \sqrt{3}, 1 - \sqrt{3})$. The first, the second, and the third invariants are 0, -6, and 4, respectively. As expected, the first invariant is zero, and the second and the third invariants equal $I_{2D} = -\frac{I_1^2}{3} + I_2 = -6$ and $I_{3D} = I_3 - \frac{I_1 I_2}{3} + \frac{2}{27} I_1^3 = 4$, respectively.

The equivalent stress equals $\sqrt{-3 I_{2D}} = \sqrt{18}$. The normal an shear octahedral stresses are $\frac{I_1}{3} = 2$ and $\sqrt{-\frac{2}{3} I_{2D}} = \frac{1}{3}\sqrt{2 I_1^2 - 6 I_2} = 2$, respectively.

The characteristic polynomial of the spherical part is $(2 - \lambda)^3 = \lambda^3 - 6\lambda^2 + 12\lambda - 8$. Invariants are $(6,12,8)$ and eigenvalues are $(2,2,2)$. The first invariant is equal to the first invariant of the original matrix, the second invariant is $\frac{1}{3} I_1^2 = 12$, and the third invariant is $\frac{1}{27} I_1^3 = 8$.

13. **The principal planes** are the planes with the normal vectors (the principal directions) on which the stress vector is purely along the normal direction and shear stresses are zero. The only three remaining normal stresses are the principal stresses and the stress tensor is diagonal. The principal stresses and directions are the eigenvalues and the eigenvectors of the stress tensor, respectively. Then

$$T_i^{(n)} = \tau_{ij}^c n_j = \lambda n_i \longrightarrow \tau_{ij}^c n_j - \lambda n_i = 0 \longrightarrow (\tau_{ij}^c - \lambda \delta_{ij}) n_j = 0 \qquad (2.12)$$

To obtain non-zero results from the last relation, we should have $det(\tau_{ij}^c - \lambda \delta_{ij}) = 0$. As explained after Equation (1.64), expansion of this determinant results in the characteristic polynomial with the tensor invariants as its coefficients and eigenvalues as its roots.

Example: Find the principal directions and principal stresses of the following plane stress field (in MPa).

$$\mathbf{D} = \begin{pmatrix} 70 & 50 \\ 50 & 70 \end{pmatrix}$$

It is easy to compute that eigenvalues of the matrix are roots of $(70 - \lambda)^2 - 50^2$, which are 20 and 120. Eigenvectors are $(1, 1)$ and $(1, -1)$ with the angle of $45°$ with respect to the x-coordinate. To find the direction of each principal stress, you have to compute the rotated stress tensor using the transformation matrix

$$\mathbf{R} = \begin{pmatrix} \cos 45° & \sin 45° \\ -\sin 45° & \cos 45° \end{pmatrix}$$

and the relation

$$\mathbf{D}' = \mathbf{R} \mathbf{D} \mathbf{R}^T = \begin{pmatrix} 120 & 0 \\ 0 & 20 \end{pmatrix}$$

14. **Compatibility condition:** Each volume cell in a continuum body is connected to other neighboring points without any gaps or overlaps. Compatibility conditions are mathematical constraints that omit the probability of the creation of such gaps or overlaps during the deformation of a body. Consequently, if a body is under stress and its deformation does not satisfy the compatibility relations, it means that the given deformation is not allowed (or the body is not a continuum).[6]

15. **The stress power** exhibits the rate of work of the surface- and volume-forces, which are connected to the stress tensor in the framework of conservation of momentum. The power of the surface force ($\mathbf{T}^{(n)} = \boldsymbol{\tau}^c \cdot \mathbf{n}$) and the external volumetric force ($\mathbf{F}$) moving with the velocity of $\mathbf{u}$ will be integrated over the surfaces and the volume, respectively

$$
\begin{aligned}
P &= \int \mathbf{T}^{(n)} \cdot \mathbf{u}\, dA + \int \mathbf{F} \cdot \mathbf{u}\, d\forall \\
&= \int (\boldsymbol{\tau}^c \cdot \mathbf{n}) \cdot \mathbf{u}\, dA + \int \mathbf{F} \cdot \mathbf{u}\, d\forall \\
&= \int \nabla \cdot (\boldsymbol{\tau}^c \cdot \mathbf{u})\, d\forall + \int \mathbf{F} \cdot \mathbf{u}\, d\forall
\end{aligned}
\tag{2.13}
$$

The last equality was obtained using the divergence theorem. Based on the Navier's equation and the identity $\nabla \cdot (\boldsymbol{\tau}^c \cdot \mathbf{u}) = (\nabla \cdot \boldsymbol{\tau}^c) \cdot \mathbf{u} + \boldsymbol{\tau}^c : \nabla \mathbf{u}$

$$
\begin{aligned}
P &= \int (\nabla \cdot \boldsymbol{\tau}^c) \cdot \mathbf{u}\, d\forall + \int \boldsymbol{\tau}^c : \nabla \mathbf{u}\, d\forall + \int \mathbf{F} \cdot \mathbf{u}\, d\forall \\
&= \int \left(\rho \frac{d\mathbf{u}}{dt} - \mathbf{F} \right) \cdot \mathbf{u}\, d\forall + \int \boldsymbol{\tau}^c : \nabla \mathbf{u}\, d\forall + \int \mathbf{F} \cdot \mathbf{u}\, d\forall \\
&= \int \rho \frac{d\mathbf{u}}{dt} \cdot \mathbf{u}\, d\forall + \int \boldsymbol{\tau}^c : \nabla \mathbf{u}\, d\forall \\
&= \frac{d}{dt} \int \frac{1}{2} \rho (\mathbf{u} \cdot \mathbf{u}) d\forall + \int \boldsymbol{\tau}^c : \nabla \mathbf{u}\, d\forall
\end{aligned}
$$

The first term in the last relation is the time rate of change of the kinetic energy. Decompose the velocity gradient tensor into its symmetric part (the rate of deformation, $\boldsymbol{\varepsilon}$) and its anti-symmetric part (the rotation tensor), and use the fact that the double dot product of a symmetric tensor ($\boldsymbol{\tau}^c$) and an anti-symmetric tensor is zero

$$
P = \frac{d}{dt} \int \frac{1}{2} \rho (\mathbf{u} \cdot \mathbf{u}) d\forall + \int \boldsymbol{\tau}^c : \boldsymbol{\varepsilon}\, d\forall
\tag{2.14}
$$

It is seen that the total power is the sum of the rate of change of kinetic energy (the bulk motion) and the stress power term, $\boldsymbol{\tau}^c : \boldsymbol{\varepsilon}$ (the work of deformations). The stress power is also linked to the dissipation function in the energy equation, which will be discussed in Section 5.2.3.

2.3 TYPES OF MATERIAL

1. **States of matter:** Continuum mechanics is the science of describing the behavior of material using the language of mathematics. Matter has seven states: liquid, solid, gas, plasma, Bose–Einstein condensate, quark-gluon plasma, and degenerate matter. In the context of the present book, we are mostly speaking about gases, liquids, and solids.

2. **General form of the constitutive relations:** Consider a typical constitutive relation for a linear material containing a property k_{ij} as a second-rank tensor,

$$
\mathbf{f} = \mathbf{k} \cdot \mathbf{d}
\tag{2.15}
$$

For example in thermal conduction, $\mathbf{f}, \mathbf{d}, \mathbf{k}$ are the heat flux vector, the temperature gradient, and negative of the second-order thermal conductivity tensor, respectively. In momentum transfer, $\mathbf{f}, \mathbf{d}, \mathbf{k}$ may be the stress tensor, the strain-rate tensor, and the fourth-order elasticity tensor, respectively.

3. **Homogenous materials** are those that have a uniform chemical composition. Hence, the properties of homogenous materials are translational invariant $\eta(\mathbf{x}) = \eta(\mathbf{x}+\mathbf{x}')$. It means that the properties such as viscosity or thermal conductivity are not an explicit function of position. Generally for a homogenous material,

$$\mathbf{k} = \begin{pmatrix} k_{11} & k_{12} & k_{13} \\ k_{21} & k_{22} & k_{23} \\ k_{31} & k_{32} & k_{33} \end{pmatrix} \tag{2.16}$$

where k_{ij} is not a function of position.

4. **Orthotropic material** has properties that only vary along three perpendicular orthogonal directions. The most famous example of orthotropic materials is wood. So, k_{ij} in the constitutive relation becomes a diagonal matrix,

$$\mathbf{k}_{orthotropic} = \begin{pmatrix} k_{11} & 0 & 0 \\ 0 & k_{22} & 0 \\ 0 & 0 & k_{33} \end{pmatrix} \tag{2.17}$$

5. **Isotropic materials** have similar properties in different directions. Hence, properties of isotropic materials are rotational invariant $\eta(\mathbf{r}) = \eta(|\mathbf{r}|)$.

$$\mathbf{k}_{isotropic} = \begin{pmatrix} k & 0 & 0 \\ 0 & k & 0 \\ 0 & 0 & k \end{pmatrix} = k\mathbf{I} \tag{2.18}$$

If the material does not obey any specific trend in terms of direction, it is called anisotropic. Examples of anisotropic substances are liquid crystals, fibrous fluids, melt of composites, molten polymers with long chains, ferrofluids in electromagnetic fields, food stuff, toothpaste, and biological fluids like blood.

6. **The scale invariance** states that the bulk properties of the matter should not be dependent on the scale and the material is self-similar.[7] When you are focusing on a continuum, the properties and the governing equations have to remain constant during zooming in. When properties of the material are size-dependent, it means that the scale invariance concept has been violated. Appearance of scale-variant film properties is a non-continuum phenomenon [1].

7. **Classification of properties:** Material properties can be classified based on the following three aspects:

   a. Physical-based classification: geometric, transport, thermodynamic, kinematic, and other properties such as surface tension coefficient and surface accommodation coefficient.

   b. Mathematical-based classification: scalar, vector, and high-rank tensor properties. For example, the speed of sound is a scalar thermodynamic property.

c. The flow-based classification in fluid mechanics: the properties of flow or fluid. For instance, density is a property of the fluid in incompressible flows, while it is a property of flow in density-varying flows.

Some common properties of matter that are used in different conservation laws or constitutive relations are as follows:

- Mass density is the most fundamental property in physics, which appears in the second law of Newton. The mass may become negative in non-continuum cases and applications such as dark fluids.
- Pressure as a negative normal stress, including static, dynamic, total (stagnation), characteristic, mechanical, thermodynamic, barometric, piezometric, vapor, osmotic, acoustic, hydraulic, gauge, and absolute pressures.
- Thermal conductivity is a transport property that appears in Fourier's law and may become negative when the anti-Fourier behavior is seen in non-continuum applications [3].
- Thermal expansion coefficient (β) represents the tendency of matter to change its volume when the temperature varies. If β is positive, the volume increases by increasing temperature and vice versa. The behavior of water near its freezing point is accompanied by the negative expansion coefficient called thermal contraction. Relations $\beta = \frac{1}{T}$, $\beta < \frac{1}{T}$, $\beta > \frac{1}{T}$ hold for perfect gas, liquids, and imperfect gases, respectively.
- Heat capacity is a thermodynamic quantity that may become negative as a non-continuum effect [4].
- The first (or the shear) and the second (or the bulk) coefficients of viscosity appear in Stokes' constitutive relation in fluid mechanics. These viscosity coefficients are counterparts of Lame's coefficients in Hook's law in elasticity theory. The first coefficient of viscosity should be positive based on the second law of thermodynamics. However, there are reports of negative [5] and zero viscosity for superfluids as a quantum non-continuum effect in fluid mechanics. Superfluids can flow without any resistance against their motion. Examples of superfluids and superfluidity are isotopes of helium (3 and 4) near cryogenic temperatures [6], applications in high-energy physics [7], and bacteria containing fluids [8].

2.4 LAGRANGIAN/EULERIAN MECHANICS

Consider a moving particle in space with the initial configuration k_0. Our final goal is to compute the deformation of that particle at time t, (k_t) and other properties of the particle. We need the time rate of change of properties since in different conservation laws, we are facing the temporal change of variables such as mass and momentum. This goal can be achieved by selecting one of Lagrangian or Eulerian approaches, and the final result should be independent of the selected viewpoint.

The Lagrangian viewpoint: Suppose that the particle starts to move from its initial position x_{0i}, along its pathline x_i. The mathematical statement of this description is $x_i = f(x_{0i}, t)$. So, x_i: the spatial coordinate is attached to the particle, and x_{0i}: the reference or the material frame is fixed at space. As a result, the configurations in the future can be mapped using f and the initial configuration of the particle.

Similar to the position vector, a general property ($\eta_{jk...}$) such as velocity, pressure, temperature, etc. in Lagrangian mechanics, can be written as a function of the initial position of the particle and time

$$\eta_{jk...} = \eta_{jk...}(x_{0i}, t) \tag{2.19}$$

Then, we can compute the time-rate of change of $\eta_{jk...}$. Since the initial position of the particle is a constant, the time-rate can be obtained by taking the derivative with respect to time,

$$\frac{d}{dt}[\eta_{jk...}(x_{0i}, t)] = \frac{\partial}{\partial t}[\eta_{jk...}(x_{0i}, t)] \tag{2.20}$$

So, the position vector is a dependent property. Velocity and acceleration equal the first and the second partial derivatives of the position vector, respectively.

The Eulerian viewpoint: In the Eulerian viewpoint, the initial position of the particle is not important; instead, the position vector x_i becomes an independent variable. So, x_i and t should be given as inputs, then we can compute other quantities at a fixed point and a specific time. In order to omit x_{0i}, we use the inverse of the mapping function f

$$x_{0i} = f^{-1}(x_i, t) \tag{2.21}$$

This inversion is only possible if the determinant of the Jacobian Matrix is non-zero. So,

$$J = \left| \frac{\partial x_i}{\partial x_{0j}} \right| = \left| \frac{\partial f_i}{\partial x_{0j}} \right| \neq 0 \tag{2.22}$$

Substitute Equation (2.21) into Equation (2.19) to omit x_{0i}, and convert the Lagrangian viewpoint to the Eulerian viewpoint,

$$\eta_{jk...}(x_{0i}, t)_L = \eta_{jk...}(f^{-1}(x_i, t), t) = \eta'_{jk...}(x_i, t)_E \tag{2.23}$$

Pay attention that the function has changed from η to η' in the Eulerian frame.

The material derivative: To compute the time-rate of change of the variable $\eta_{jk...}$, we have to use the chain-rule; since the position vector x_i is a function of time.

$$\frac{d}{dt}[\eta_{jk...}(x_{0i}, t)]_L = \frac{\partial}{\partial t}[\eta'_{jk...}(x_i, t)] + \frac{\partial}{\partial x_m}[\eta'_{jk...}(x_i, t)]\frac{dx_m}{dt} = \frac{D}{Dt}[\eta'_{jk...}(x_i, t)]_E$$
$$\tag{2.24}$$

The operator $\frac{D}{Dt}$ is just defined to simplify the notations. The first and the second terms in the material derivative operator containing the partial derivatives with respect to time and position are called the local term and the advection (or convection) term, respectively. The other names for the material derivative operator are the substantial derivative, the total derivative, and the convection derivative. The integral form of the material derivative is connected to the Reynolds transport theorem (RTT).

The ALE formulation: If the properties of the material are expressed with respect to neither the moving frame attached to the particle nor the fixed frame, but with respect to a moving frame with an arbitrary velocity, the new description is called the arbitrary Eulerian–Lagrangian (ALE) formulation. The ALE viewpoint can be used in problems involving the free-surface or the moving boundaries. The only modification needed to derive the ALE version of the material derivative is that the velocity in the convection term should be written with respect to the moving frame.

$$\frac{D}{Dt}[\eta'_{jk\ldots}(x_i,t)] = \frac{\partial}{\partial t}[\eta'_{jk\ldots}(x_i,t)] + \frac{\partial}{\partial x_m}[\eta'_{jk\ldots}(x_i,t)]\frac{dx_m}{dt}\Big|_R \qquad (2.25)$$

where $\frac{dx_m}{dt}\big|_R$ is the relative velocity of the particle. It is obvious that if the reference frame is fixed (the Eulerian frame), the relative velocity equals the absolute velocity, and the material derivative operator appears. If the frame moves with the velocity of the particle (the Lagrangian derivative), the relative velocity is zero, the advection term vanishes, and the material derivative reduces to the partial derivative with respect to time.

Example: Consider the pathlines of particles in the following Lagrangian transient velocity field.

$$\begin{aligned} x &= x_0 + at \\ y &= y_0 \\ z &= z_0 + cy_0^3 t^2 \end{aligned} \qquad (2.26)$$

Compute the Jacobian matrix and its determinant, then calculate the displacement field in the Eulerian frame.

Solution: The determinant of the Jacobian matrix of the above-mentioned transformation reads

$$J = det\left(\frac{\partial x_i}{\partial x_{0j}}\right) = det\begin{pmatrix} \frac{\partial x}{\partial x_0} & \frac{\partial x}{\partial y_0} & \frac{\partial x}{\partial z_0} \\ \frac{\partial y}{\partial x_0} & \frac{\partial y}{\partial y_0} & \frac{\partial y}{\partial z_0} \\ \frac{\partial z}{\partial x_0} & \frac{\partial z}{\partial y_0} & \frac{\partial z}{\partial z_0} \end{pmatrix} = det\begin{pmatrix} 1 & 0 & 0 \\ 0 & 1 & 0 \\ 0 & 3cy_0^2 t^2 & 1 \end{pmatrix} \neq 0$$

$$(2.27)$$

The non-zero Jacobian determinant implies that the unique inverse of the mapping $x_i = f(x_{0i},t)$ exists: $x_{i0} = f^{-1}(x_i,t)$. The determinant of the Jacobian matrix equals the ratio of volumes in transformed and original frames, and hence, it should be non-zero. This is a condition for detecting admissible deformations. The inverse of the above-mentioned mapping expressing the Eulerian frame is $x_0 = x - at$, $y_0 = y$, and $z_0 = z - cy^3 t^2$.

2.5 EXERCISES

2.1 If eigenvalues and eigenvectors of a given transformation ($\mathbf{A}$) is known, compute the eigenvalues and eigenvectors of a new transformation which is obtained by applying the transformation five successive times $\mathbf{A}^5$.

2.2 Prove that the advection part of the material derivative can be written as the sum of the Lamb vector and half of the gradient of $\mathbf{u} \cdot \mathbf{u}$.

2.3 Find the traction vector and the normal stress at the point p located on the plane $2x_1 - 2x_2 - x_3 = 0$, if the stress tensor at that point is

$$\begin{pmatrix} 1 & 2 & 3 \\ 2 & 4 & 6 \\ 3 & 6 & 1 \end{pmatrix} \tag{P2.1}$$

2.4 Compute the curl of a shear flow with the velocity field $(ky, 0, 0)^T$. Then write down the given velocity as a sum of $\frac{k}{2}(y, x)^T$ and $\frac{k}{2}(y, -x)^T$. Compute the gradients of two new velocity fields by decomposing into symmetric and anti-symmetric parts.

2.5 If axes of coordinates are principle directions of the stress tensor, compute the normal stress on a plain with unit normal vector $\frac{1}{\sqrt{3}}(1, 1, -1)^T$ as a function of the first invariant of the stress tensor.

2.6 If the velocity field in the Lagrangian viewpoint is $u_i = \alpha(X_i + 1)$ where α is a positive constant, compute the velocity components in Eulerian description.

2.7 If the stress tensor equals

$$\begin{pmatrix} x_1 & -x_2 & f(x_1) \\ -x_2 & x_3 & 0 \\ f(x_1) & 0 & 2x_3 \end{pmatrix}$$

and the body is in equilibrium without any external body force, compute the function $f(x_1)$.

2.8 If the product of eigenvalues of the following tensor equals 120, find the value or values of x.

$$\begin{pmatrix} 6 & x & x \\ x & 8 & 4 \\ x & 4 & 9 \end{pmatrix}$$

2.9 Obtain the time-rate of change of temperature field $T = x_1 + tx_2$ when the displacement in Lagrangian viewpoint is

$$\begin{aligned} x_1 &= X_1 + AtX_2 \\ x_2 &= X_2 - AtX_1 \\ x_3 &= X_3 \end{aligned}$$

2.10 Obtain magnitudes of σ_1 and σ_2, if the principle stresses of the following stress tensor are 4, 3, -2 MPa.

$$\begin{pmatrix} \sigma_1 & -3 & 0 \\ -3 & \sigma_2 & 0 \\ 0 & 0 & 3 \end{pmatrix} MPa$$

2.11   A displacement field in Lagrangian viewpoint is given.

$$x_1 = \lambda_1 X_1$$
$$x_2 = -\lambda_3 X_3$$
$$x_3 = \lambda_2 X_2$$

Compute the Jacobian of the transformation.

2.12   Compute unknowns a, b, c, if the stress vector components on the octahedral plane for the following stress tensor are zero.

$$\begin{pmatrix} 1 & a & b \\ a & 1 & c \\ b & c & 1 \end{pmatrix}$$

2.13   Principle stresses of a stress field are 10, 20, 30 MPa. Compute the equivalent stress.

NOTES

1. Based on atomism philosophy
2. Since, entropy is not conserved, this equation is called the entropy balance. Likewise, the energy equation in cosmic scales may be called the energy balance.
3. A similar concept exists in special relativity theorem called the Lorentz invariance.
4. Similar to the appearance of two pink surfaces when you are slicing a watermelon.
5. Postulate or axiom is a statement that is supposed to be true without any specific proof.
6. There is another restriction on the stability of solids called the Drucker stability postulate that is a relation between the stress and the strain increments $d\tau^c : d\varepsilon \geq 0$
7. Fractals and turbulent flows are examples of self-similar geometrical and physical structures, respectively.

3 Newtonian and Non-Newtonian Fluids

The continuity equation (conservation of mass) and constitutive relations for fluids and fluid-like materials will be investigated in this chapter. First, we have to derive the components of the strain-rate tensor as kinematic relations based on velocity derivatives. Then, we need some constitutive relations to compute the stress tensor from the stain-rate tensor.

3.1 THE STRAIN-RATE TENSOR

Figure 3.1 illustrates the motion and deformation of a two-dimensional rectangular element in the xy plane inside a velocity field during the interval dt. We are trying to calculate changes in the position and shape of different parts of the continuum media due to the motion of sides and edges of a differential representative element. The motion of points A to D of the element within a specific velocity field will result in rotation, translation, distortion, or extension of that element (Q3.1).

The lengths of different parts of the element as a function of the symbols shown in the figure are

$$A'O' = dy + \frac{\partial v}{\partial y} dy dt$$

$$A'O = dx + \frac{\partial u}{\partial x} dx dt$$

$$OB' = \frac{\partial v}{\partial x} dx dt$$

$$O'C' = \frac{\partial u}{\partial y} dy dt$$

$$(3.1)$$

The element illustrated in Figure 3.1 experiences linear and rotational motions, along with linear and angular deformations.

1. **Rigid body linear motion** can be computed using the velocity field. This type of motion disappears when $\mathbf{u} = \mathbf{0}$.
2. **Rigid body rotation** can be calculated using the mean change of angles of two sides of the element. The counter-clockwise rotation is positive and vice versa. For an element in the xy plane, the mean angular rotation about the z-axis is

$$d\Omega_z = \frac{1}{2}(da - db) \qquad (3.2)$$

DOI: 10.1201/9781032719405-3

48

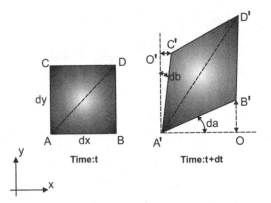

Figure 3.1 Schematic of a continuum fluid element with the dilatation, the angular deformation, the rigid body rotation, and the linear motion.

in which

$$da = \lim_{dt \to 0} \left(\tan^{-1} \frac{\frac{\partial v}{\partial x} dx dt}{dx + \frac{\partial u}{\partial x} dx dt} \right) = \frac{\partial v}{\partial x} dt$$

$$db = \lim_{dt \to 0} \left(\tan^{-1} \frac{\frac{\partial u}{\partial y} dy dt}{dy + \frac{\partial v}{\partial y} dy dt} \right) = \frac{\partial u}{\partial y} dt \tag{3.3}$$

This type of motion is absent when $da = db$ (distortion without rotation). Two other components Ω_x and Ω_y can be computed in a similar manner. So, the angular velocity components in three directions equal

$$\frac{d\Omega_z}{dt} = \frac{1}{2} \left(\frac{\partial v}{\partial x} - \frac{\partial u}{\partial y} \right)$$

$$\frac{d\Omega_x}{dt} = \frac{1}{2} \left(\frac{\partial w}{\partial y} - \frac{\partial v}{\partial z} \right) \rightarrow \frac{d\Omega}{dt} = \frac{\nabla \times u}{2} = \frac{\omega}{2}$$

$$\frac{d\Omega_y}{dt} = \frac{1}{2} \left(\frac{\partial u}{\partial z} - \frac{\partial w}{\partial x} \right) \tag{3.4}$$

The curl of the velocity field is called vorticity ($\boldsymbol{\omega}$). When the flow field is curl-free, this type of motion vanishes.

3. **The shear or the distortional strain-rate** for a two-dimensional element in the xy plane is computed by the mean reduction of the angle between two sides

of the element, which were initially perpendicular,

$$\varepsilon_{xy} \times dt = \frac{1}{2}(da + db) = \frac{1}{2}\left(\frac{\partial u}{\partial y} + \frac{\partial v}{\partial x}\right)dt \rightarrow \varepsilon_{xy} = \frac{1}{2}\left(\frac{\partial u}{\partial y} + \frac{\partial v}{\partial x}\right)$$

$$\varepsilon_{yz} = \frac{1}{2}\left(\frac{\partial w}{\partial y} + \frac{\partial v}{\partial z}\right)$$

$$\varepsilon_{zx} = \frac{1}{2}\left(\frac{\partial u}{\partial z} + \frac{\partial w}{\partial x}\right)$$

$$(3.5)$$

where $\varepsilon_{xy} = \varepsilon_{yx}$, $\varepsilon_{yz} = \varepsilon_{zy}$, $\varepsilon_{xz} = \varepsilon_{zx}$ are the off-diagonal elements of the strain-rate tensor. This type of deformation vanishes when $da = -db$ (rotation without distortion). From the procedure of derivation of the strain-rate components, it is obvious that the strain-rate tensor is symmetric $\varepsilon_{ij} = \varepsilon_{ji}$.

4. **The dilation**[1] **or the extensional strain-rate** is calculated by the increases in the distance between two parallel sides of the element. For instance, in the x-direction

$$\varepsilon_{xx}dt = \frac{(dx + \frac{\partial u}{\partial x}dxdt) - dx}{dx} = \frac{\partial u}{\partial x}dt \Rightarrow \varepsilon_{xx} = \frac{\partial u}{\partial x} \qquad (3.6)$$

Similarly, for two other directions

$$\varepsilon_{yy} = \frac{\partial v}{\partial y}, \quad \varepsilon_{zz} = \frac{\partial w}{\partial z} \qquad (3.7)$$

The strain-rate tensor: Combining the mentioned motions and deformations in a tensor, the final form of the strain-rate tensor reads

$$\begin{pmatrix} \varepsilon_{xx} & \varepsilon_{xy} & \varepsilon_{xz} \\ \varepsilon_{yx} & \varepsilon_{yy} & \varepsilon_{yz} \\ \varepsilon_{zx} & \varepsilon_{zy} & \varepsilon_{zz} \end{pmatrix} = \begin{pmatrix} \frac{\partial u}{\partial x} & \frac{1}{2}\left(\frac{\partial u}{\partial y} + \frac{\partial v}{\partial x}\right) & \frac{1}{2}\left(\frac{\partial u}{\partial z} + \frac{\partial w}{\partial x}\right) \\ \frac{1}{2}\left(\frac{\partial v}{\partial x} + \frac{\partial u}{\partial y}\right) & \frac{\partial v}{\partial y} & \frac{1}{2}\left(\frac{\partial v}{\partial z} + \frac{\partial w}{\partial y}\right) \\ \frac{1}{2}\left(\frac{\partial w}{\partial x} + \frac{\partial u}{\partial z}\right) & \frac{1}{2}\left(\frac{\partial w}{\partial y} + \frac{\partial v}{\partial z}\right) & \frac{\partial w}{\partial z} \end{pmatrix} \qquad (3.8)$$

The strain-rate tensor is obtained from neither a conservation law nor a constitutive relation. It is the fruit of a kinematic analysis of an element of the material.

The trace of the strain-rate tensor or its first invariant is equal to the divergence of the velocity field. The trace of the strain-rate tensor equals the sum of three dilations and exhibits the total rate of change of volume of the element per unit volume. The volume change in x, y, z directions are $(\frac{\partial u}{\partial x}dxdt)dydz$, $(\frac{\partial v}{\partial y}dydt)dxdz$, $(\frac{\partial w}{\partial z}dzdt)dxdy$, respectively. Then the total rate of change of volume per volume of the differential element $(\frac{1}{d\forall}\frac{Dd\forall}{Dt})$ is

$$\frac{1}{d\forall}\frac{Dd\forall}{Dt} = \varepsilon_{ii} = \nabla \cdot \mathbf{u} \qquad (3.9)$$

where the term ε_{ii} obeys the summation convention. The flow is incompressible when the strain-rate tensor is traceless, $\varepsilon_{ii} = \nabla \cdot \mathbf{u} = 0$. If the trace of the strain-rate tensor is a positive or negative number, the element experiences an increase or a reduction in volume, respectively.

When one of the terms in $\frac{\partial u_i}{\partial x_i}$ is positive, and two others are negative, the case is called the uniaxial elongational flow, like what occurs in vortex filament stretching. The case of one negative and two positive terms is called the biaxial elongational flow. If one term is zero, and two others are opposite in sign with equal magnitudes, the case is called the planar elongational flow. This case is similar to what happens when a layer of fluid confined between two fixed constraints is stretched, and its thickness reduces.

The strain-rate decomposition: Since the flow element simultaneously rotates while linearly/angularly deforms, the velocity gradient tensor should be asymmetric. In order to determine the rate of these two effects, we split the velocity gradient tensor into the symmetric and antisymmetric parts as discussed in Equation (1.51) (Q3.2).

$$\frac{\partial u_i}{\partial x_j} = \underbrace{\varepsilon_{ij}}_{Strain\,rate} + \underbrace{\frac{d\Omega_{ij}}{dt}}_{Angular\,velocity} \tag{3.10}$$

where the first term in the right-hand side is the symmetric strain-rate tensor, and the second term is the anti-symmetric rotation tensor. The rotation tensor is related to the vorticity vector.

3.2 CONSERVATION OF MASS: THE CONTINUITY EQUATION

The conservation of mass can be derived from the material derivative of a differential mass which should be zero,

$$\frac{D(dm)}{Dt} = \frac{D(\rho d\forall)}{Dt} = \rho \frac{D(d\forall)}{Dt} + d\forall \frac{D\rho}{Dt} = 0 \tag{3.11}$$

Dividing both sides by the volume of the element and using Equation (3.9), the continuity equation reads

$$\frac{D\rho}{Dt} + \rho \nabla \cdot \mathbf{u} = 0 \rightarrow \frac{\partial \rho}{\partial t} + \nabla \cdot (\rho \mathbf{u}) = 0 \tag{3.12}$$

$\frac{\partial \rho}{\partial t} + \nabla \cdot (\rho \mathbf{u}) = 0$ is called the conservation form of the mass conservation law. If the density is a constant, the velocity field becomes a divergence-free field. If the flow field is steady but compressible, the quantity $\rho \mathbf{u}$ becomes divergenceless.

The divergence operator: We noticed that in incompressible flows, the velocity field is divergenceless. In order to understand the physical meaning of the divergence operator, consider a cubic element of fluid. If two parallel sides are stationary or they keep their distance during their motion, the volume of the element does not change.

In order to produce a volume change in direction x, the velocity component in x-direction should change in that direction ($\partial u/\partial x \neq 0$). If the velocity gradient in the x-direction is positive, it means that the walls are moving away, and the volume of the element increases and vice versa. In conclusion, if we want to explore a quantity proportional to the total volume change of an element, we have to sum the velocity gradients in three directions. This way, the divergence operator appears. Consider that zero volume change does not necessarily happen when the velocity components are zero. We may have positive $\partial u/\partial x$ and negative $\partial v/\partial y$ with the same magnitudes. Hence, the velocity changes while the volume of the element does not.

An alternative form: Using simple algebra

$$-\frac{1}{\rho}\frac{D\rho}{Dt} = \nabla \cdot \mathbf{u} \rightarrow \frac{D\ln v}{Dt} = \nabla \cdot \mathbf{u} \tag{3.13}$$

where v is the specific volume. Since $v = 0$, and ∞ produce singularities in Equation (3.13), the density of a continuum cannot become either zero or infinity.

Molecular diffusion of mass: Generally speaking, the mass can be transferred by the molecular diffusion mechanism, which was ignored in Equation (3.12). The left-hand side of Equation (3.12) denotes the bulk convective mass transport. The zero in the right-hand side of the equation is an estimation of the molecular mass diffusion and can be substituted by $-\nabla \cdot \dot{\mathbf{m}}_D$ where (Q3.3)

$$\dot{\mathbf{m}}_D = -D\nabla\rho - \frac{D\rho}{2T}\nabla T \tag{3.14}$$

where $\dot{\mathbf{m}}_D$ is the diffusion mass flow rate per unit area. The first and the second terms correspond to Fick's mass diffusion and Soret's thermal diffusion terms, respectively. There is also another term related to the mass transfer from pressure difference. This term has been ignored here. D and $\frac{D\rho}{2T}$ are the mass diffusivity and Soret's diffusivity coefficients, respectively.

If the mass diffusion velocity is defined as $U_{Di} = \frac{\dot{m}_{Di}}{\rho}$, the total velocity can be written as $U_{Ti} = U_{Di} + u_i$. Then, the total form of the continuity equation is written as

$$\frac{D\rho}{Dt} + \rho\nabla \cdot \mathbf{U}_T = 0 \tag{3.15}$$

which has a structure similar to the pure-convection mass conservation law. The diffusion of mass originating from the Brownian motion of particles weakens discontinuities such as density jump across the interfaces between two fluids.

Example: Use the conservation of mass to simplify $\mathbf{I}$ that is the inertial force in the equation of motion of a fluid in the conservation version. Show that $\mathbf{I}$ reduces to the material derivative operator.

$$
\begin{aligned}
\mathbf{I} &= \frac{\partial(\rho\mathbf{u})}{\partial t} + \nabla \cdot (\rho\mathbf{u}\otimes\mathbf{u}) = \rho\frac{\partial\mathbf{u}}{\partial t} + \mathbf{u}\frac{\partial\rho}{\partial t} + \nabla \cdot \begin{pmatrix} (\rho u)u & (\rho u)v & (\rho u)w \\ (\rho v)u & (\rho v)v & (\rho v)w \\ (\rho w)u & (\rho w)v & (\rho w)w \end{pmatrix} \\
&= \rho\frac{\partial\mathbf{u}}{\partial t} + \mathbf{u}\frac{\partial\rho}{\partial t} + \begin{pmatrix} \frac{\partial(\rho u)u}{\partial x} + \frac{\partial(\rho u)v}{\partial y} + \frac{\partial(\rho u)w}{\partial z} \\ \frac{\partial(\rho v)u}{\partial x} + \frac{\partial(\rho v)v}{\partial y} + \frac{\partial(\rho v)w}{\partial z} \\ \frac{\partial(\rho w)u}{\partial x} + \frac{\partial(\rho w)v}{\partial y} + \frac{\partial(\rho w)w}{\partial z} \end{pmatrix}
\end{aligned}
$$

$$\tag{3.16}$$

For instance, the x-component of the inertial force in cartesian coordinates is

$$\rightarrow I_x = \left(\rho \frac{\partial u}{\partial t} + u \frac{\partial \rho}{\partial t} \right)$$

$$+ \left(u \frac{\partial(\rho u)}{\partial x} + (\rho u)\frac{\partial u}{\partial x} + u \frac{\partial(\rho v)}{\partial y} + (\rho v)\frac{\partial u}{\partial y} + u \frac{\partial(\rho w)}{\partial z} + (\rho w)\frac{\partial u}{\partial z} \right)$$

$$= u \left(\underbrace{\frac{\partial \rho}{\partial t} + \nabla \cdot (\rho \mathbf{u})}_{0} \right) + \rho \left(\frac{\partial u}{\partial t} + u\frac{\partial u}{\partial x} + v\frac{\partial u}{\partial y} + w\frac{\partial u}{\partial z} \right)$$

$$= \rho \left(\frac{\partial u}{\partial t} + u\frac{\partial u}{\partial x} + v\frac{\partial u}{\partial y} + w\frac{\partial u}{\partial z} \right)$$

The underlined terms have been combined to form the divergence of the product of the velocity and the density field. The zero term is substituted from the conservation of mass equation. The final result is

$$\mathbf{I} = \rho \left[\frac{\partial \mathbf{u}}{\partial t} + (\mathbf{u} \cdot \nabla)\mathbf{u} \right] = \rho \frac{D\mathbf{u}}{Dt}$$

Example: Show that the time-derivative of the volume integral of the product of a scalar and the density equals the volume integral of the product of the density and the material derivative of that quantity.

Solution: Use Leibniz's formula and apply the divergence theorem,

$$\int_S (\phi \mathbf{u}_s) \cdot \hat{n} dS = \int_\forall \nabla \cdot (\phi \mathbf{u}_s) d\forall \tag{3.17}$$

for a scalar $\phi = \rho f$ where ρ is the density, to obtain,

$$\frac{d}{dt} \int_{\forall(t)} \rho f d\forall = \int_{\forall(t)} \left[\frac{\partial(\rho f)}{\partial t} + \nabla \cdot (\rho f \mathbf{u}_s) \right] d\forall$$

$$= \int_{\forall(t)} \left[\rho \frac{\partial f}{\partial t} + \rho \mathbf{u}_s \cdot \nabla f + \underbrace{f\frac{\partial \rho}{\partial t} + f \nabla \cdot (\rho \mathbf{u}_s)}_{0} \right] d\forall$$

$$= \int_{\forall(t)} \rho \frac{Df}{Dt} d\forall$$

The conservation of mass ($\frac{\partial \rho}{\partial t} + \nabla \cdot (\rho \mathbf{u}_s) = 0$) has been used. So, in this case, the order of integration and differentiation can be changed using the material derivative.

3.2.1 STREAM FUNCTION VS. POTENTIAL FUNCTION

The definition of stream function is a result of the conservation of mass. However, there is another quantity called the potential function, which is apparently similar to but fundamentally different from the stream function.

Based on Helmholtz's decomposition, a vector like **u** can be decomposed into a divergenceless part and a curl-free part

$$\mathbf{u} = \nabla \times \mathbf{A} + \nabla f \tag{3.18}$$

Potential function: The divergence of curl of any vector is zero; so, the first term in the RHS (the rotational part) is divergence-free or solenoidal. Taking the divergence of both sides of Equation (3.18) yields

$$\nabla^2 f = \nabla \cdot \mathbf{u} \tag{3.19}$$

If the velocity field is incompressible, its divergence is zero, and the governing equation to find f will be the Laplace equation. f for irrotational flows ($\nabla \times \mathbf{u} = \nabla \times \nabla f = 0$) is called the velocity potential (ϕ). For a three-dimensional curl-free velocity field, the velocity can be defined as the gradient of the potential function,

$$\mathbf{u} = \nabla \phi \rightarrow u = \frac{\partial \phi}{\partial x}, v = \frac{\partial \phi}{\partial y}, w = \frac{\partial \phi}{\partial z} \tag{3.20}$$

The velocity potential is a scalar function and is the direct result of the irrotationality of the flow field.

Stream function: The curl of gradient of any scalar is zero; so, the second term in the RHS of Equation (3.18) or the irrotational part is curl-free. The gauge condition is that the vector **A** should be divergenceless. Taking the curl of both sides of Equation (3.18) yields

$$\nabla \times \mathbf{u} = \nabla \times \nabla \times \mathbf{A} = \nabla(\nabla \cdot \mathbf{A}) - \nabla^2 \mathbf{A} \rightarrow \nabla^2 \mathbf{A} = -\nabla \times \mathbf{u} \tag{3.21}$$

If the flow field is irrotational, its curl is zero, and the governing equation to find **A** will be the Laplace equation. For incompressible flows, $f \equiv 0$, $\mathbf{A} \equiv \boldsymbol{\psi}$, and $\boldsymbol{\psi}$ is the three-dimensional form of the stream vector.

It is common to use the two-dimensional form of $\boldsymbol{\psi} = \psi \mathbf{e}_z$, in which ψ is a scalar. The following relation is the definition of the stream function for a two-dimensional incompressible flow

$$\mathbf{u} = \nabla \times (\psi \mathbf{e}_z) \rightarrow u = \frac{\partial \psi}{\partial y}, v = -\frac{\partial \psi}{\partial x} \tag{3.22}$$

Properties: Important properties of these two scalar kinematic functions are listed as follows:

1. Orthogonality of ψ-constant and ϕ-constant lines,

$$d\psi = \frac{\partial \psi}{\partial x} dx + \frac{\partial \psi}{\partial y} dy$$

$$d\phi = \frac{\partial \phi}{\partial x} dx + \frac{\partial \phi}{\partial y} dy$$

So,

$$\frac{dy}{dx}\Big|_{\psi-cte.} = \frac{v}{u}$$

$$\frac{dy}{dx}\Big|_{\phi-cte.} = -\frac{u}{v}$$

which proves that the two lines are perpendicular.

2. The iso-stream-function lines are streamlines. ds is the differential element vector parallel to the velocity or tangential to the streamline, then

$$\mathbf{u} \times d\mathbf{s} = \begin{vmatrix} \mathbf{i} & \mathbf{j} & \mathbf{k} \\ u & v & w \\ dx & dy & dz \end{vmatrix}$$

$$= 0 \rightarrow \mathbf{i}(vdz - wdy) + \mathbf{j}(wdx - udz) + \mathbf{k}(udy - vdx) = 0 \rightarrow$$

$$\frac{dz}{dy} = \frac{w}{v}, \quad \frac{dz}{dx} = \frac{w}{u}, \quad \frac{dy}{dx} = \frac{v}{u} \tag{3.23}$$

These are the definitions of streamlines.

3. The compressible form of stream function (ψ') for 2D steady flow reads

$$\frac{\partial \rho}{\partial t} + \nabla \cdot (\rho \mathbf{u}) \;=\; 0 \rightarrow \frac{\partial(\rho u)}{\partial x} + \frac{\partial(\rho v)}{\partial y} = 0 \rightarrow$$

$$\rho u \;=\; \frac{\partial \psi'}{\partial y}, \rho v = -\frac{\partial \psi'}{\partial x}. \tag{3.24}$$

The dimension of ψ' is different from its incompressible counterpart. It should be noted that this compressible form is just useful when the variation of density with temperature or salinity is known. For high-Mach number compressible flows in which the density is a function of pressure and temperature, the compressible definition of the stream function is almost useless.

4. The definition of stream function or potential function helps us reduce two unknowns (the velocity components) to one unknown.

5. The relation between stream function and vorticity for two-dimensional cases in the xy plane is

$$\omega_z = \left(\frac{\partial v}{\partial x} - \frac{\partial u}{\partial y} \right) = -\frac{\partial^2 \psi}{\partial x^2} - \frac{\partial^2 \psi}{\partial y^2} \rightarrow \omega_z = -\nabla^2 \psi \tag{3.25}$$

6. The two-dimensional volumetric flow rate confined between to streamlines per unit length normal to the page remains constant. As shown in Figure 3.2a, any arbitrary cross-section between two plane streamlines can be decomposed into two components A_x and A_y.

$$Q_{1-2} = Q_{1-0} + Q_{0-2} = \int_1^0 u\,dy - \int_0^2 v\,dx = \int_1^2 (u\,dy - v\,dx) = \int_1^2 d\psi = \psi_2 - \psi_1 \tag{3.26}$$

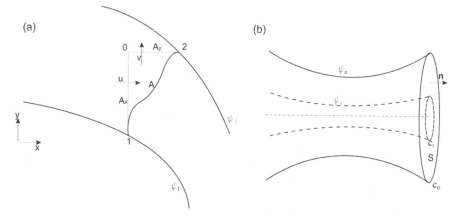

Figure 3.2 Configuration of (a) area and velocity components between two streamlines, (b) axisymmetric streamlines forming a streamtube.

If $\psi_2 > \psi_1$, the flow streams from left to right and vice versa. If the compressible stream function is replaced, the mass flow rate per unit length normal to the page will be obtained.

7. The definition of stream function in curvilinear coordinates, including the polar and spherical coordinates, is also important in some applications (Q3.4).

• **The first cylindrical form, the axisymmetric version, the Stokes stream function**: The cylindrical form of the mass conservation law reads

$$\frac{\partial \rho}{\partial t} + \frac{1}{r}\frac{\partial(r\rho u_r)}{\partial r} + \frac{1}{r}\frac{\partial(\rho u_\theta)}{\partial \theta} + \frac{\partial(\rho u_z)}{\partial z} = 0 \qquad (3.27)$$

The incompressible steady form is

$$\frac{1}{r}\frac{\partial(r u_r)}{\partial r} + \frac{1}{r}\frac{\partial u_\theta}{\partial \theta} + \frac{\partial u_z}{\partial z} = 0 \qquad (3.28)$$

The Stokes stream function can be demonstrated when the second term related to the azimuthal coordinate vanishes

$$u_r = -\frac{1}{r}\frac{\partial \psi}{\partial z}$$

$$u_z = \frac{1}{r}\frac{\partial \psi}{\partial r} \rightarrow \mathbf{u} = \nabla \times \left(\frac{1}{r}\psi \mathbf{e}_\theta\right)$$

The dimension of the Stokes stream function is L^3/t, and the volumetric flow rate can be computed using $Q = 2\pi(\psi_o - \psi_i)$. To prove this relation, consider two concentric streamtubes. As shown in Figure 3.2b, the magnitudes of the stream functions on axisymmetric streamtubes are ψ_o and ψ_i, and the normal vector ($\mathbf{n}$) of the area confined between two tubes (c_o and c_i) have been specified on the figure.

$$Q = \int_s \mathbf{u} \cdot d\mathbf{s} = \int_s \nabla \times \left(\frac{1}{r} \psi \mathbf{e}_\theta \right) \cdot d\mathbf{s} \qquad (3.29)$$

Using Stokes' integral theorem

$$Q = \oint_{c_o} \frac{1}{r} \psi \mathbf{e}_\theta \cdot d\mathbf{l} + \oint_{c_i} \frac{1}{r} \psi \mathbf{e}_\theta \cdot d\mathbf{l} \qquad (3.30)$$

We know that the value of the stream function is constant on stream-lines crossing the c-curve,

$$Q = \psi_o \oint_{c_o} \frac{1}{r} \mathbf{e}_\theta \cdot d\mathbf{l} + \psi_i \oint_{c_i} \frac{1}{r} \mathbf{e}_\theta \cdot d\mathbf{l} \qquad (3.31)$$

Substitute the vector $d\mathbf{l} = dr \mathbf{e}_r + r d\theta \mathbf{e}_\theta$

$$\mathbf{e}_\theta \cdot d\mathbf{l} = \mathbf{e}_\theta \cdot (dr \mathbf{e}_r + r d\theta \mathbf{e}_\theta) = 0 + r d\theta = r d\theta \rightarrow$$

$$Q = \psi_o \int_0^{2\pi} d\theta + \psi_i \int_{2\pi}^0 d\theta = 2\pi(\psi_o - \psi_i) \qquad (3.32)$$

Again, if the value of the stream function on the outer stream tube is greater than the inner one, the net flow rate will be from left to right.

- **The second cylindrical form, the polar form**: If the third term in Equation (3.28) related to the z-coordinate is zero, the second cylindrical form of the stream function can be written as

$$\begin{aligned} u_r &= \frac{1}{r} \frac{\partial \psi}{\partial \theta} \\ u_z &= -\frac{\partial \psi}{\partial r} \end{aligned} \quad \rightarrow \mathbf{u} = \nabla \times (\psi \mathbf{e}_z) \qquad (3.33)$$

The dimension of the polar stream function is proportional to the volumetric flow rate per unit length normal to the page L^2/t, and the plane volumetric flow rate in this case can be computed using $Q = \psi_2 - \psi_1$.

- **The spherical form**: The steady form of the conservation of mass in spherical coordinates is

$$\frac{1}{R} \frac{\partial (R^2 u_R)}{\partial R} + \frac{1}{\sin\theta} \frac{\partial u_\phi}{\partial \phi} + \frac{1}{\sin\theta} \frac{\partial (u_\theta \sin\theta)}{\partial \theta} = 0 \qquad (3.34)$$

After neglecting the second term, the spherical form of the stream function $\mathbf{u} = \nabla \times \left(\frac{\psi \mathbf{e}_\phi}{R \sin\theta} \right)$ reads

$$\begin{aligned} u_R &= \frac{1}{R^2 \sin\theta} \frac{\partial \psi}{\partial \theta} \\ u_\theta &= -\frac{1}{R \sin\theta} \frac{\partial \psi}{\partial R} \end{aligned} \qquad (3.35)$$

where ϕ and θ are the azimuthal and the coaltitude (polar) angles, respectively.

Legendre potential function Φ is defined by applying the Legendre transformation to the potential function, $\Phi = xu + yv + zw - \phi$ such that $d\Phi = xdu + ydv + zdw$. Hence, $x = \frac{\partial \Phi}{\partial u}, y = \frac{\partial \Phi}{\partial v}, z = \frac{\partial \Phi}{\partial w}$. The Chaplygin's equation has been developed based on the Legendre potential function to model the two-dimensional transonic compressible flows,

$$\frac{\partial^2 \Phi}{\partial \theta^2} + \frac{v^2}{1 - M^2} \frac{\partial^2 \Phi}{\partial v^2} + v \frac{\partial \Phi}{\partial v} = 0 \qquad (3.36)$$

where the independent variables are θ and v. v is the magnitude of the velocity vector in two-dimensional space, and θ is the angle between the velocity vector and the u component.

3.3 CONSTITUTIVE RELATIONS

In this section, a transformation to map the deformation rate tensor (as input) to the stress tensor (as output) will be computed. Based on the rule of ranks, this mapping should be a fourth-order tensor called the elasticity tensor.

3.3.1 THE STRESS TENSOR (Q3.5)

Stokesian fluid: A relation between the stress tensor and the strain-rate tensor for fluids was derived by George Stokes, similar to Hook's law in elasticity. Stokes made the following fundamental postulates.

1. The stress tensor is a continuous function of the strain-rate tensor and thermodynamic state of the system. So, the stress tensor is independent of the rigid body rotation (the rotation tensor) introduced in Equation (3.10).
2. The stress tensor is not explicitly a function of position, and the fluid is homogeneous. The dependence of the stress tensor on position is via the strain-rate tensor as a function of velocity and material properties.
3. The fluid is isotropic, and hence, different directions have no privilege. There is no priority for any specific direction. If the stress tensor along a specific coordinate becomes a diagonal tensor, the strain-rate tensor along that direction should also become another diagonal tensor. Stresses create deformations in their direction. So, we may relate the normal, and the shear stresses to the linear and angular deformation rates, respectively. In other words, if the shear stresses are zero, then the shear deformation rates have to be zero as well. In some cases, such as powders, exerting a normal stress in a direction applies shear that leads to the rotational motion of the granular particles.
4. At the limiting case of hydrostatic fluid, the stress tensor should be equal to its spherical part $-p\mathbf{I}$, where $\mathbf{I}$ is the second-order identity tensor. So, we can define a viscous stress tensor τ_{ij} by excluding $-p\mathbf{I}$. The viscous stress tensor denotes the deviatoric part of the stress tensor that should be symmetric only if the stress tensor is symmetric.

A general non-elastic fluid, which obeys the conditions mentioned above, is called the Stokesian fluid or the Reiner–Rivlin fluid. The constitutive relation of the

Stokesian fluid is

$$\tau_{ij}^c = -p\delta_{ij} + \tau_{ij} = (-p+L)\delta_{ij} + 2M\varepsilon_{ij} + 2N\varepsilon_{ik}\varepsilon_{kj} \tag{3.37}$$

which is a nonlinear relation that satisfies the fourth assumption regarding the hydrostatic limit. In the case of Newtonian fluid as a special case of the Stokesian fluid, the last nonlinear term is zero ($N = 0$), $L = \lambda\nabla\cdot\mathbf{u}$, $M = \mu$ in which λ is the second coefficient of viscosity.

Newtonian fluid: The Newtonian constitutive relation is the linear version of the constitutive relation of the Stokesian fluid, and relates the viscous stress tensor (τ_{ij}) and the deformation rate tensor by a fourth-order tensor,

$$\tau_{ij} = K_{ijkm}\varepsilon_{km} \tag{3.38}$$

Based on the rule of ranks, the order of the connecting tensor of two second-order tensors should be 4. Since the fluid is isotropic, the general form of a fourth-order isotropic tensor should be selected for A_{ijkm} containing three position-independent scalars B, C, D,

$$A_{ijkm} = B\delta_{ik}\delta_{jm} + C\delta_{im}\delta_{jk} + D\delta_{ij}\delta_{km} \tag{3.39}$$

Since the stress tensor is usually symmetric, the relation should not change by swapping i and j, and consequently, $B = C$. Then, the stress tensor is

$$
\begin{aligned}
\tau_{ij} &= C(\delta_{ik}\delta_{jm} + \delta_{im}\delta_{jk})\varepsilon_{km} + D\delta_{ij}\delta_{km}\varepsilon_{km} \\
&= C(\varepsilon_{ij} + \varepsilon_{ji}) + D\delta_{ij}\varepsilon_{kk} \\
&= 2C\varepsilon_{ij} + D\delta_{ij}\varepsilon_{kk}
\end{aligned}
\tag{3.40}
$$

The last relation is obtained based on the symmetry of the strain-rate tensor. Constants C and D are independent material properties which are called the first and the second coefficients of viscosity, μ and λ, respectively. They are not explicit functions of position. A new term $-p\delta_{ij}$ also should be added to justify the hydrostatic limit.

$$\tau_{ij}^c = -p\delta_{ij} + 2\mu\varepsilon_{ij} + \lambda\delta_{ij}\varepsilon_{kk} \tag{3.41}$$

Substituting the strain-rate tensor, the final form of the stress tensor for a Newtonian fluid reads

$$\tau_{ij}^c = -p\delta_{ij} + \mu\left(\frac{\partial u_i}{\partial x_j} + \frac{\partial u_j}{\partial x_i}\right) + \lambda\delta_{ij}\frac{\partial u_k}{\partial x_k} \tag{3.42}$$

The stress tensor, excluding the pressure term, is called the viscous stress tensor $\boldsymbol{\tau}$, which is the deviatoric part of the stress tensor. For incompressible flows, the last term as well as λ in Equation (3.42) vanish (Q3.6).

For incompressible flows, a new term $c\delta_{ij}$ could be added and combined with the pressure term due to their similar mathematical structure. Since we use the divergence of the stress tensor in the conservation laws, you may add any constant to the pressure without affecting the result. In incompressible flows, only the change of pressure is important,[2] and you are free to shift the magnitude of pressure by a constant value. However, for compressible cases, the pressure appears in the equation of state. So, just a specific scalar $\lambda \delta_{ij} \frac{\partial u_k}{\partial x_k}$ can be added to the pressure term.

The final form of the stress tensor for a Newtonian fluid reads

$$
\begin{pmatrix}
-p + \lambda \nabla \cdot \mathbf{u} + 2\mu \frac{\partial u}{\partial x} & \mu \left(\frac{\partial u}{\partial y} + \frac{\partial v}{\partial x} \right) & \mu \left(\frac{\partial u}{\partial z} + \frac{\partial w}{\partial x} \right) \\
\mu \left(\frac{\partial v}{\partial x} + \frac{\partial u}{\partial y} \right) & -p + \lambda \nabla \cdot \mathbf{u} + 2\mu \frac{\partial v}{\partial y} & \mu \left(\frac{\partial v}{\partial z} + \frac{\partial w}{\partial y} \right) \\
\mu \left(\frac{\partial w}{\partial x} + \frac{\partial u}{\partial z} \right) & \mu \left(\frac{\partial w}{\partial y} + \frac{\partial v}{\partial z} \right) & -p + \lambda \nabla \cdot \mathbf{u} + 2\mu \frac{\partial w}{\partial z}
\end{pmatrix}
\tag{3.43}
$$

As it is obvious, the normal stresses have three origins: the hydrostatic pressure $(-p)$ which is the same in three directions, the resistance of fluid against total volume change $(\lambda \nabla \cdot \mathbf{u})$ that is again similar for three directions, and resistance of fluid against shear due to volume change in one direction, $(2\mu \frac{\partial u}{\partial x}$ along the x-direction for instance).

Molecular diffusion: Similar to the mass conservation law, the momentum transfer due to mass diffusion has to be added to the stress tensor to obtain

$$
\tau_{ij}^{c,D} = -p\delta_{ij} + \mu \left(\frac{\partial u_i}{\partial x_j} + \frac{\partial u_j}{\partial x_i} \right) + \lambda \delta_{ij} \frac{\partial u_k}{\partial x_k} - \left[\dot{m}_{Di} u_j + \dot{m}_{Dj} u_i + \frac{\lambda}{\mu} \delta_{ij} \dot{m}_{Dk} u_k \right]
\tag{3.44}
$$

The three new terms correspond to the diffusion mechanism, which are usually ignored compared to the bulk momentum transfer. The momentum transfer due to mass diffusion weakens the discontinuities in parallel-to-interface velocity component across the interface between two fluids, such as the slip-line and the vortex sheet.

Types of pressure: Pressure is a scalar quantity[3] acting as normal stress. Besides defining concepts such as pressure loss, pressure drop, pressure difference, gauge pressure, and absolute pressure, there exist other types of pressure in diverse applications as follows.

1. The hydrostatic pressure equals the product of the density, the gravitational acceleration, and the fluid height above a point $(\rho g z)$.
2. The barometric pressure is another name for atmospheric pressure.
3. The static pressure is a type of pressure measured independent of the velocity of the fluid.
4. The dynamic pressure is a combination of the magnitude of the velocity vector and the density of the fluid $(\rho \frac{u^2}{2})$.
5. The piezometric pressure is the sum of the hydrostatic pressure and the static pressure $(\rho g z + p)$.
6. The vapor pressure at a specific temperate is the pressure at which the phase-change (cavitation) would occur. This name is also used as the partial pressure of water vapor in the air that is equal to the product of the relative humidity and the saturation pressure.
7. The osmotic pressure. Suppose that we have filled two arms of a U-shaped pipe with two solutions with dissimilar concentrations separated by a permeable

membrane at the central cross section of the tube. It is seen that the height of the fluids inside the two arms differ by a certain height. The osmotic pressure is the product of the fluid's specific weight and the difference in heights of the fluid in two arms.

8. The stagnation/ram/total pressure is the sum of the static and the dynamic pressures $(p + \rho \frac{u^2}{2})$.

9. The viscous pressure is defined in low-Reynolds creeping flows as a reference pressure $(\mu \frac{u}{L})$.

10. The relative pressure which is defined to perform computations for isentropic process of ideal gases when specific heats are temperature-dependent $(C_1 \exp \frac{s^\circ}{R})$.

11. The reduced pressure equals the ratio of thermodynamic pressure and critical pressure of the gas, which is helpful to define the degree of the imperfectness of gases.

12. The characteristics pressure is defined in compressible flows as the pressure of a gas when its Mach number adiabatically is increased or decreased to reach unity.

13. The mechanical pressure (p^*) is derived from the stress tensor as the negative of arithmetic mean of the normal stresses,

$$
\begin{aligned}
p^* &= -\frac{\tau_{ii}^c}{3} = -\frac{I_1}{3} \\
&= -\frac{1}{3}\left[\left(-p + 2\mu\frac{\partial u}{\partial x} + \lambda \nabla \cdot \mathbf{u}\right) + \left(-p + 2\mu\frac{\partial v}{\partial y} + \lambda \nabla \cdot \mathbf{u}\right) \right. \\
&\quad \left. + \left(-p + 2\mu\frac{\partial w}{\partial z} + \lambda \nabla \cdot \mathbf{u}\right)\right] \\
&= -\frac{1}{3}\left[-3p + 2\mu\left(\frac{\partial u}{\partial x} + \frac{\partial v}{\partial y} + \frac{\partial w}{\partial z}\right) + 3\lambda \nabla \cdot \mathbf{u}\right] \\
&= -\frac{1}{3}\left(-3p + 2\mu\nabla \cdot \mathbf{u} + 3\lambda \nabla \cdot \mathbf{u}\right) \\
&= p - \left(\lambda + \frac{2}{3}\mu\right)\nabla \cdot \mathbf{u} = p - K\nabla \cdot \mathbf{u} = p + K\frac{1}{\rho}\frac{D\rho}{Dt}
\end{aligned}
\tag{3.45}
$$

where, hereinafter, K will be called the Stokesity. The last relation has been obtained using the continuity equation.

14. The thermodynamic pressure obeys the two-property rule in thermodynamics. It is a function of two other intensive properties (for a simple compressible substance). Pressure is a hydraulic quantity in incompressible flows and a thermodynamic quantity in compressible flows.

A question arises here. When are the mechanical pressure and the thermodynamic pressure identical? The following items are answers based on Equation (3.45),

- The fluid is hydrostatic.
- The flow is divergenceless.

- K=0. This condition implies that the rate of change of the volume of an element does not contribute to pressure. This contradiction may be solved by defining $K\nabla \cdot \mathbf{u} = -p$, and supposing K to be a function of density, $K(\rho)$. Kozachok (2019) indicated that the Stokesity could be obtained directly from the Cauchy stress tensor, instead of the deviatoric part, and pointed out that the pressure is related to the trace of the stress tensor. This way, we can prevent facing the definitions of the thermodynamic pressure and the mechanical pressure [10].

15. Acoustic pressure demonstrates the deviation of pressure from the atmospheric surrounding pressure. The origin of such deviation is the propagation of sound waves with the speed of sound. The product of the acoustic pressure and velocity is called sound intensity. A logarithmic scale of the acoustic pressure compared to a reference leads to the definition of sound pressure levels.

Stokes hypothesis: George Stokes talked about the second coefficient of viscosity, or the volume viscosity or the dilation viscosity, or the bulk viscosity. The value of the second coefficient of viscosity was almost unknown. He simply supposed that the mechanical and the thermodynamic pressures are equal and used this statement to find a relation to close equations. Consequently, the Stokes hypothesis is equivalent to zero Stokesity,

$$K = \lambda + \frac{2}{3}\mu = 0 \tag{3.46}$$

After more than 150 years of this simple assumption, we still do not have a strong theory about this hypothesis. A more strange fact is that despite of the simplicity of this assumption, the results obtained from the solution of the conservation laws derived based on this relation are remarkably exact. Buresti (2015) [11] proposed that instead of zero Stokesity assumption, we may assume that the absolute value of $K\nabla \cdot \mathbf{u}$ is negligible in comparison to the thermodynamic pressure $|K\nabla \cdot \mathbf{u}| \ll p$.

From the kinetic theory of gases, the mechanical pressure is proportional to the translational energy of particles. But, the thermodynamic pressure accounts for the total energy of particles, including vibrational, rotational, and intermolecular bonds. The Stokes hypothesis demonstrates the difference between these molecular energies. Accordingly, the Stokes hypothesis is valid if the gas particles contain only the translational energy (monatomic dilute gases). Then, the mechanical and the thermodynamic pressures become equal [12].

The Stokes hypothesis is invalid when the medium is not a continuum, or the thermodynamic equilibrium fails, in small-scale phenomena such as shock waves in which a large density-gradient appears along a distance in the order of microns, in short-time-scale phenomena such as high-intensity ultrasound with frequencies even much higher than the human threshold of hearing, and during the sound wave absorption or attenuation[4] [13].

Second coefficient of viscosity: μ is related to the dissipation of energy due to shear.[5] However, λ represents the dissipation mechanism against volume change without shear. Based on previous discussions, the second coefficient of viscosity appears when an element is expanded or compressed. It means that there is extra resistance due to the compressibility of flow.

Consider the normal and shear stresses in the x direction

$$\tau_{xx}^c = -p + \lambda \nabla \cdot \mathbf{u} + 2\mu \frac{\partial u}{\partial x}$$

$$\tau_{xy}^c = \mu \left(\frac{\partial u}{\partial y} + \frac{\partial v}{\partial x} \right)$$

The normal stress along the x-direction (τ_{xx}^c) consists of three parts. The first part is related to the static pressure, and the second part containing the bulk viscosity, demonstrates the dissipative resistance stemming from the total volume change. The third term containing the shear viscosity originates from the shear due to a part of the total volume change in the x-direction ($\frac{\partial u}{\partial x}$).

If an element of a fluid with negative λ expands, a compressive negative normal stress appears due to the second term. Since μ is a positive variable (referring to the second law of thermodynamics), the normal stress in the x-direction contributed from the third term is positive when the element expands along the x-direction and vice versa.

The value of bulk viscosity depends on the frequency of the sound wave that is being attenuated or absorbed. Due to its frequency-dependence nature, we cannot say that the bulk viscosity is a thermodynamic quantity. The value of the second coefficient of viscosity in some cases is a positive constant and even can be greater than μ, for water as a liquid is 2.4 cp, for carbon-sulfide is greater than 200 cp, for fluids such as Benzene may become hundreds or even thousands times greater than the shear viscosity [13, 14]. Despite attempts to measure the second coefficient of viscosity, such as what Geoffrey Taylor did, we need a more coherent theory to estimate the magnitude of the second coefficient of viscosity for liquids and polyatomic gases.

3.3.2 NON-NEWTONIAN FLUIDS AND POWDERS

First, it should be noted that most non-Newtonian fluids in common applications are liquids.[6] Non-Newtonian gases are not among the working fluids in normal engineering situations. So, we may assume the incompressibility condition for non-Newtonian fluids except in cases such as water hammer or working under high pressure differences. This assumption neglects the density fluctuation in liquids and ignores the compressibility effects in liquids, such as what happens in water hammer. Consequently, a constitutive relation will be developed using an incompressible platform similar to Newtonian fluids (Q3.7).

It is so common to introduce the apparent viscosity to relate the viscous stress tensor to temperature, pressure, the shear rate parameter ε_I, time, and the yield stress,

$$\tau_{ij} = 2\mu_A(T, P, \varepsilon_I, t, \tau_c, \dots)\varepsilon_{ij} \tag{3.47}$$

where τ_c is the yield stress, $\frac{1}{2}\varepsilon_I^2 = \boldsymbol{\varepsilon} : \boldsymbol{\varepsilon} = tr(\boldsymbol{\varepsilon}^2) = \varepsilon_{ij}\varepsilon_{ji}$ is the second invariant of the strain-rate tensor. In two-dimensional cases, ε_I is simplified to

$$\varepsilon_I = \left\{ 2\left[\left(\frac{\partial u}{\partial x}\right)^2 + \left(\frac{\partial v}{\partial y}\right)^2 \right] + \left[\frac{\partial v}{\partial x} + \frac{\partial u}{\partial y}\right]^2 \right\}^{0.5} \tag{3.48}$$

and for one-dimensional boundary layer flows in which the velocity component in the x-direction only varies with y,

$$\varepsilon_I = |\frac{\partial u}{\partial y}| \tag{3.49}$$

If the apparent viscosity increases by enhancing the shear rate, it is called the shear-thickening fluid, and the reversed case is called the shear-thinning case. If the apparent viscosity increases over time, it is called rheopectic, and the reversed case is thixotropic. Based on the definitions mentioned above, different types of fluids are listed here.

1. **Newtonian fluids** for which the viscosity is only a function of temperature and pressure. The viscosity of liquids decreases with increasing temperature. The viscosity of dilute gases increases with temperature. The viscosity of both liquids and gases may slightly increase with pressure. The most famous relation which presents a relation for the dependency of the dynamic viscosity on temperature is the celebrated Sutherland's relation,

$$\frac{\mu}{\mu_0} = \left(\frac{T}{T_0}\right)\frac{T_0 + S}{T + S} \tag{3.50}$$

   where the subscript "0" refers to a known reference condition, and S is Sutherland's constant, which equals 110 K for air.
   Another relation is Maxwell's power-law equation

$$\frac{\mu}{\mu_0} = \left(\frac{T}{T_0}\right)^n \tag{3.51}$$

   where n varies from 0.6 to 1.2 for different gases. $n = 0.7$ is a good suggestion [13].

2. **The time-independent/dependent models**
   - The power-law approximation is the most famous model for a time-independent fluid.

$$\mu_A = C\varepsilon_I^{n-1} \tag{3.52}$$

   where C and n are the consistency coefficient and the power-law index, respectively, and can be obtained from curve fitting.
   - The Carreau–Yasuda–Cross model for the Carreau fluid again as a time-independent fluid

$$\mu_A = \mu_\infty + (\mu_0 - \mu_\infty)[1 + t_0^m \varepsilon_I^m]^{\frac{n-1}{m}} \tag{3.53}$$

μ_0 and μ_∞ are viscosities at zero and infinite shear rates, respectively, t_0 is the relaxation time, and $m = 2$. Other parameters may be obtained from experimental data.

Tiu–Boger's model and Weltman's model are examples of time-dependent constitutive relations.

3. **Fluids with the yield stress:** For these fluids, the deformation appears just above a specific shear stress called the yield stress (τ_c).

- The apparent viscosity for the Bingham fluid with plastic deformation reads

$$\mu_A = \mu_0 + \frac{\tau_c}{\varepsilon_I}; \quad |\tau| > \tau_c \tag{3.54}$$

otherwise, the apparent viscosity is infinite. Infinite apparent viscosity means that a tremendous force is needed to create a deformation in the fluid. The apparent viscosity for a one-dimensional flow is simplified to $\mu_A = \mu_0 + \frac{\tau_c}{|\frac{\partial u}{\partial y}|}$, and hence, the stress–strain-rate relation is $\tau_{xy} = \mu_0 |\frac{\partial u}{\partial y}| + \tau_c$. Imagine a fluid with the yield stress, inside a vertical cylindrical shell. Try to raise the shell. The weight of the fluid appears as a shear stress on walls. If the fluid attaches to the shell, it means that the shear stress on walls due to the weight of the fluid is smaller than the yield stress of the fluid.

- The Herschel–Bulkley fluid is another type of non-Newtonian fluid with the yield stress,

$$\mu_A = k\varepsilon_I^{n-1} + \frac{\tau_c}{\varepsilon_I}; \quad |\tau| > \tau_c \tag{3.55}$$

otherwise, the apparent viscosity is μ_0, which may be a large number similar to the Bingham fluid model. n is the flow index and k is the consistency constant.

Other visco-plastic models are the Casson model and its extension, the Quemada model. The Casson model is mainly suggested for blood flow simulations.

4. **Powder flows:** The Schaeffer's law is a well-known model to describe the fluid-like behavior of powders,

$$\mu_A = \frac{p\sin(\phi)}{\varepsilon_I}; \quad |\tau| > \tau_c \tag{3.56}$$

Schaeffer's law is similar to the Bingham fluid's constitutive relations except that the pressure and a new constant ϕ are added in the yield stress term, and μ_0 has been eliminated. So, powders can be regarded as a sub-category of plastic material. In this model, a negative pressure may produce negative apparent viscosity. This new behavior in powder dynamics is called dilatancy.

$\tau_c = p\sin(\phi)$ can be computed based on the von Mises yield condition in plasticity, in which $\sin(\phi)$ is a constant of the material. ϕ represents the angle of internal friction of the material. The presence of p, in this relation, implies that the pressure as a normal stress affects the shear stresses. Hence, the viscous term is pressure-dependent [16].

Pay attention that for general powder flows, the principal directions of the stress and the strain-rate tensors are not parallel. So, some postulates must be considered in Schaeffer's theory to deal with this problem. Also, some non-Schaefferian and particle-based Lagrangian models have been presented. Such models try to may eliminate a part of deficiencies in Schaeffer's law [17].

5. **Viscoelastic fluids:** From continuum mechanics viewpoint, different materials can be classified into three categories: pure fluids: the stress tensor is related to the strain-rate tensor, and the strain-rate tensor is connected to the velocity; pure solids: the stress tensor is related to the strain tensor, and the strain tensor is connected to the displacement (Section 6.1); viscoelastic materials: the stress or the stress-rate tensor may be related to both the strain and the strain-rate tensors.

   • The Maxwell's model is

$$\frac{E}{\eta}\tau_{ij} + \dot{\tau}_{ij} = E\dot{\varepsilon}_{ij}^r \tag{3.57}$$

   where ε^r is the strain tensor, and "dot" represents the time derivative. E and η are Young's modulus (stiffness) and viscosity, respectively.
   • The Voigt–Kelvin model is

$$\varepsilon_{ij} + \frac{E}{\eta}\varepsilon_{ij}^r = \frac{\tau_{ij}}{\eta} \tag{3.58}$$

   There are other contributions such as fractional form of Maxwell's model [18], stochastic model of viscoelastic material [19], and viscoelastic model for heterogenous media [20].

6. **The second-order fluid:** The constitutive relation for this kind of non-Newtonian fluid contains a second-order term,

$$\boldsymbol{\tau} = \lambda \nabla \cdot \mathbf{u} + 2\mu\boldsymbol{\varepsilon} + 2\alpha_1 \mathbf{A} + 4\alpha_2 \boldsymbol{\varepsilon}^2 \tag{3.59}$$

where $\alpha_1 < 0$ and α_2 are the normal stress modules. $\mathbf{A}$ is the Rivlin–Ericksen tensor defined as

$$\mathbf{A} = \frac{D\boldsymbol{\varepsilon}}{Dt} + \boldsymbol{\varepsilon}\nabla\mathbf{u} + (\nabla\mathbf{u})^T\boldsymbol{\varepsilon} \tag{3.60}$$

7. **The complex fluid:** The complex fluid simultaneously shows fluid-like and solid-like behaviors. The viscous stress tensor for a polymeric complex flow is written as the sum of the solvent tensor and the polymer tensor [21]

$$\boldsymbol{\tau} = \boldsymbol{\tau}_s + \boldsymbol{\tau}_p \tag{3.61}$$

The solvent part may obey the Newtonian constitutive relation, and $\boldsymbol{\tau}_p$ for a complex fluid such as the Oldroyd-B fluid is defined as

$$\boldsymbol{\tau}_p = \frac{\mu_p k}{\mu_p + \mu_s}(\mathbf{c} - \mathbf{I}) \tag{3.62}$$

$\mathbf{c}$ is the conformation tensor

$$\frac{\partial \mathbf{c}}{\partial t} + \mathbf{u} \cdot \nabla \mathbf{c} - \mathbf{c} \cdot \nabla \mathbf{u} = \nabla \mathbf{u}^T \cdot \mathbf{c} + \frac{k}{\mu_s + \mu_p}(\mathbf{I} - \mathbf{c}) \tag{3.63}$$

8. **The second-gradient fluid:** The first version of the constitutive relation for the gradient fluid has been presented by Korteweg [22]. In gradient fluids, the stress–strain rate relation contains some new gradient terms. However, research about the gradient materials is mostly related to solids. The Korteweg's model includes the density gradient as follows,

$$
\begin{aligned}
\boldsymbol{\tau}^c &= \boldsymbol{\tau}_{dissipative} + \boldsymbol{\tau}_{elastic} \\
&= [2\mu\boldsymbol{\varepsilon} + \lambda(\nabla \cdot \mathbf{u})\mathbf{I}] + [-p\mathbf{I} + (\alpha_0|\nabla\rho|^2 + \alpha_1\nabla^2\rho)\mathbf{I} + \alpha_3(\nabla\rho \otimes \nabla\rho)]
\end{aligned}
\tag{3.64}
$$

where $\boldsymbol{\varepsilon}$, $\nabla\rho \otimes \nabla\rho$, α_0, α_1, α_2 are the symmetric part of the velocity gradient tensor, the Korteweg tensor, and the material properties depending on density, respectively.

NOTES

1. Dilatation is also correct.
2. A well-known exception is cavitation, in which the pressure magnitude is also important to be higher than the vapor pressure to prevent cavitation.
3. Pressure can be treated as a second-order tensor in applications such as statistical mechanics and inhomogeneous fluids [9].
4. It is recommended to study the Stokes' law of attenuation.
5. In uniaxial elongational flows, the ratio of normal stress difference to the stretching rate is measured by Trouton viscosity or extensional viscosity. For Newtonian liquids, the Trouton viscosity equals three times the shear viscosity (Trouton's ratio).
6. Non-Newtonian behavior is seen in gases at low-temperatures related to quantum fluid mechanics [15].

4 Navier–Stokes Equations

In this chapter, we will derive the Navier–Stokes equations (conservation of momentum) for fluids and fluid-like materials. We will substitute the stress tensor discussed in the previous chapter in the momentum conservation law to find the full Navier–Stokes equations and the alternative and limiting forms. The dimensionless form of the laws, the boundary conditions, and the conservation of angular momentum are other sections of the present chapter.

4.1 CONSERVATION OF LINEAR MOMENTUM

Density plays a central role in the second law of Newton. Fluid flows can be classified based on the variation of density,

1. Incompressible flow, $\rho = c$. The flow is divergence-free, and λ disappears.
2. p-compressibility, $\rho = f_1(p, T)$ happens in high-Mach number flows. The flow is not divergence-free, and λ remains in formulations.
3. w-compressibility, $\rho = f_2(p)$ happens in applications such as water hammer or oil flow under huge pressure-rise inside a pump.
4. T-compressibility, $\rho = f_3(T)$ may appear even for liquids such as natural convection in seawater, the thermosyphon phenomenon, and the formation of a breeze in the air.
5. s-compressibility, $\rho = f_4(T, s)$. Here, the density-change is related to the temperature and salinity of the seawater (Q4.1).

4.1.1 THE NAVIER–STOKES EQUATION IN PRIMITIVE VARIABLES (4.2)

Starting from the second law of Newton and dividing both sides by the volume of a differential element,

$$\Sigma \mathbf{F} = \rho \mathbf{a} \tag{4.1}$$

where $\mathbf{F}$ is the force per unit volume or the force density and equals the sum of line-loading (surface tension), surface, and volume forces: $\mathbf{F}_{ST} + \mathbf{F}_s + \mathbf{F}_B$.

The surface forces stemming from the shear and the normal stresses per unit volume exerted on the surfaces of a differential element of fluid equals $\mathbf{F}_s = \nabla \cdot \boldsymbol{\tau}$. As an example, let us derive the x-component of the surface force (f_{sx}) and the surface

DOI: 10.1201/9781032719405-4

force density (F_{sx}) using Taylor's theorem,

$$
\begin{aligned}
df_{sx} &= \left(\left[\tau_{xx}^c + \frac{\partial \tau_{xx}^c}{\partial x}dx\right] - \tau_{xx}^c\right)dydz + \left(\left[\tau_{yx}^c + \frac{\partial \tau_{yx}^c}{\partial y}dy\right] - \tau_{yx}^c\right)dxdz \\
&\quad + \left(\left[\tau_{zx}^c + \frac{\partial \tau_{zx}^c}{\partial z}dz\right] - \tau_{zx}^c\right)dxdy \\
&= \left(\frac{\partial \tau_{xx}^c}{\partial x} + \frac{\partial \tau_{yx}^c}{\partial y} + \frac{\partial \tau_{zx}^c}{\partial z}\right)d\forall \\
\rightarrow F_{sx} &= \frac{df_x}{d\forall} = \frac{\partial \tau_{xx}^c}{\partial x} + \frac{\partial \tau_{yx}^c}{\partial y} + \frac{\partial \tau_{zx}^c}{\partial z} \\
\rightarrow \mathbf{F}_s &= \nabla \cdot \boldsymbol{\tau}^c \tag{4.2}
\end{aligned}
$$

Substituting the acceleration using the material derivative of the velocity vector and the force vector as the sum of the body force term and the surface forces, we finally obtain the full Navier–Stokes equation,

$$
\rho\left(\underbrace{\frac{D\mathbf{u}}{Dt}}_{1} + \underbrace{\mathbf{a}_R}_{2}\right) = \underbrace{-\nabla p}_{3} + \underbrace{\nabla \cdot \boldsymbol{\tau}}_{4} + \underbrace{\nabla \cdot \boldsymbol{\tau}_D}_{5} + \underbrace{\mathbf{F}_g}_{6} + \underbrace{\mathbf{F}_{Lo}}_{7} + \underbrace{\mathbf{F}_{EO}}_{8}
$$
$$
+ \underbrace{\mathbf{F}_{ST}}_{9} + \underbrace{\mathbf{F}_T}_{10} + \underbrace{\mathbf{F}_{FC}}_{11} + \underbrace{\mathbf{F}_{TP}}_{12} + \underbrace{\mathbf{F}_{St}}_{13} \tag{4.3}
$$

All terms in Equation (4.3) have the dimension of force per unit volume. Different terms in the Navier–Stokes equation are introduced one-by-one following the numbers devoted to each term within the equation (Q4.3).

1. **The acceleration term:** The acceleration vector equals the material derivative of the velocity field,

$$
\frac{D\mathbf{u}}{Dt} = \frac{\partial \mathbf{u}}{\partial t} + (\mathbf{u}_r \cdot \nabla)\mathbf{u} \tag{4.4}
$$

The fractional form: For some special forms of non-local cases, non-integer values for derivatives may be used to construct the time-fractional form of the Navier–Stokes equation in the framework of fractional calculus [23]. More general details about the non-local and the weakly non-local fluid mechanics can be found in Eringen (1974) [24] and Van and Fulop (2006) [25], respectively.

The ALE formulation: In classical fluid mechanics, the Eulerian viewpoint is used in which $\mathbf{u}_r = \mathbf{u}$. In fully Lagrangian approaches $\mathbf{u}_r = 0$, the nonlinear advection terms disappear. If the fluid is streaming into a domain with deforming boundaries, the velocity vector that is the multiplier in the material derivative operator should be computed relative to the velocity of the domain, such as the dynamic/moving mesh $\mathbf{u}_r = \mathbf{u} - \mathbf{u}_{mesh}$.

This approach is called the arbitrary Lagrangian–Eulerian (ALE) formulation and is frequently used in the simulation of moving boundary problems. In computational methods for moving boundary flows, $\mathbf{u}_r$ is equal to the velocity of the fluid minus the velocity of the moving mesh [26]. A similar trend is also valid for the material derivative term in the energy equation [27–33] and the conservation of mass

$$\frac{D\rho}{Dt} + \rho \nabla \cdot \mathbf{u} = 0 \rightarrow \left[\frac{\partial \rho}{\partial t} + (\mathbf{u}_r \cdot \nabla)\rho\right] + \rho \nabla \cdot \mathbf{u} = 0 \qquad (4.5)$$

The ALE concept is similar to how we deal with a moving control volume in undergraduate fluid mechanics. Similarly, for a moving control volume, the velocities in the Reynolds transport theorem (RTT) operator have to be written relative to the moving coordinates.

2. **The non-inertial terms:** If the coordinates are attached to the control volume under study, we have to add relative accelerations to the conservation of linear momentum. The relative accelerations with respect to the non-inertial coordinates are the linear, the Coriolis, the centripetal, and the angular accelerations

$$\mathbf{a}_R = \mathbf{a}_l + 2\boldsymbol{\Omega} \times \mathbf{u} + \boldsymbol{\Omega} \times (\boldsymbol{\Omega} \times \mathbf{x}) + \dot{\boldsymbol{\Omega}} \times \mathbf{x} \qquad (4.6)$$

The Coriolis acceleration appears in geophysical flows around the earth. The important point is that the Coriolis acceleration due to the earth's rotation becomes important in large-scale, slowly moving problems in which the Rossby number goes to zero. So, small-scale problems such as the rotation of swirling water in a sink or even tornadoes are not massively affected by the Coriolis acceleration. Three important applications in which the Coriolis effect appears are the stream of fluid inside pump rotating impellers, the Ekman spiral that appears in ocean flows near poles and the Taylor column that is seen in a rotating fluid perturbed by a solid body.

3. **The pressure term:** Using the obtained constitutive relation, the surface force per unit volume (equal to the divergence of the stress tensor) reads

$$\begin{aligned}
\nabla \cdot \tau^c &= \frac{\partial}{\partial x_j}\left[\delta_{ij}\left(\lambda u_{k,k} - p\right) + \mu\left(u_{i,j} + u_{j,i}\right)\right] \\
&= -\nabla p + \frac{\partial}{\partial x_j}\left[\lambda \nabla \cdot \mathbf{u}\delta_{ij} + \mu\left(u_{i,j} + u_{j,i}\right)\right] \\
&= -\nabla p + \nabla(\lambda \nabla \cdot \mathbf{u}) + \nabla \cdot \left[\mu(\nabla \mathbf{u} + \nabla \mathbf{u}^T)\right] \\
&= -\nabla p + \mu \nabla^2 \mathbf{u} + (\mu + \lambda)\nabla(\nabla \cdot \mathbf{u}) \qquad (4.7)
\end{aligned}$$

The first term in the equation above stems from the hydrostatic part of the stress tensor and is called the pressure force (per unit volume). The negative sign indicates that the pressure force is exerted from the high-pressure region toward the low-pressure zone. It should be noted that the final relation is obtained for a constant-property flow using the following identity.

$$\nabla \cdot (\mu \nabla \mathbf{u}^T) = \mu \frac{\partial}{\partial x_j}\left(\frac{\partial u_j}{\partial x_i}\right) = \mu \frac{\partial}{\partial x_i}\left(\frac{\partial u_j}{\partial x_j}\right) = \mu \nabla(\nabla \cdot \mathbf{u})$$

4. **The viscous term:** Two last terms in Equation (4.7) originate from the deviatoric part of the stress tensor and are called the viscous terms. Such expressions can be simplified under different conditions, which will be discussed in the following items.

Incompressible form with constant viscosity: In this case, the viscous terms reduce to the Laplacian operator

$$
\begin{aligned}
\nabla \cdot \tau^c &= -\nabla p + \mu \nabla \cdot (\nabla \mathbf{u}) \\
&= -\nabla p + \mu \nabla^2 \mathbf{u}
\end{aligned} \tag{4.8}
$$

For the sake of clarity, we repeat calculations in cartesian coordinates in the x-direction,

$$
\begin{aligned}
\nabla \cdot \tau^c|_x &= -\frac{\partial p}{\partial x} + \frac{\partial}{\partial x}\left[2\mu\frac{\partial u}{\partial x} + \lambda \nabla \cdot \mathbf{u}\right] \\
&\quad + \frac{\partial}{\partial y}\left[\mu\left(\frac{\partial u}{\partial y} + \frac{\partial v}{\partial x}\right)\right] + \frac{\partial}{\partial z}\left[\mu\left(\frac{\partial u}{\partial z} + \frac{\partial w}{\partial x}\right)\right] \\
&= -\frac{\partial p}{\partial x} + \mu\left[\frac{\partial^2 u}{\partial x^2} + \frac{\partial^2 u}{\partial y^2} + \frac{\partial^2 u}{\partial z^2}\right] + \mu\frac{\partial}{\partial x}\underbrace{\left[\frac{\partial u}{\partial x} + \frac{\partial v}{\partial y} + \frac{\partial w}{\partial z}\right]}_{0} \\
&= -\frac{\partial p}{\partial x} + \mu\nabla^2 u
\end{aligned}
$$

The fractional form: The fractional form of the viscous term can be written as

$$
\nabla \cdot \tau = \mu\frac{\partial^\alpha u}{\partial |y|^\alpha} \tag{4.9}
$$

where the fractional derivative with order $1 < \alpha \leq 2$ is computed based on the Riesz fractional derivative operator. A more comprehensive study of the fractional flows will be presented in Section 6.10.1. Some special considerations should be paid to the dimensional consistency of terms and constants.

Variable-viscosity fluids: The viscosity coefficient can be a function of temperature or pressure, even in Newtonian fluids. The viscous stress tensor of non-Newtonian fluids, or variable-viscosity Newtonian fluids may be a function of temperature, time, shear rate or yield stress. So, the divergence of the stress tensor should be replaced by using the apparent viscosity concept $\mu_A(T, t, \varepsilon_I, \tau_c, ...)$.

Anisotropic form: In this case, the viscosity should be written as a second-order tensor. This way, the viscosity in all relations acts like a matrix, and a dot symbol should be added to the viscous term between the viscosity tensor and other quantities. For a Stokesian fluid, the viscosity tensor reads

$$
\mu_{ij} = M\left(\delta_{ij} + \varepsilon_{ij}\frac{N}{M}\right) \tag{4.10}
$$

where constants M and N have been introduced in Equation (3.37). The viscosity tensor reduces to $\mu\mathbf{I}$ for isotropic Newtonian fluids. Anisotropic form of the viscous force for incompressible flows in which the viscosity is different in various directions is

$$\nabla \cdot \boldsymbol{\tau} = \mu_{x,y}(\nabla^2)_{x,y}\mathbf{u} + \mu_z\frac{\partial^2\mathbf{u}}{\partial z^2} \qquad (4.11)$$

where $\mu_{x,y}$ and μ_z are the directional viscosities. $(\nabla^2)_{x,y}$ is the directional Laplacian operator that only contains the second-order partial derivatives in the x and y directions. So, the diffusion of momentum differs in different directions [35].

Turbulence averaging: Incompressible form of the viscous term after conducting the time-averaging yields

$$\nabla \cdot \boldsymbol{\tau} = \frac{\partial}{\partial x_j}\left(\mu\frac{\partial \overline{u_i}}{\partial x_j} - \rho\overline{u'_i u'_j}\right) \qquad (4.12)$$

New terms are called Reynolds stress tensor. More details will be presented in Section 6.6.

5. **Molecular diffusion:** This part is obtained from the divergence of additional terms corresponding to the mass diffusion in the stress–strain rate constitutive relation (Equation 3.44) appear.

6. **The gravity force:** If we neglect the temperature dependence of the density as it happens in free convection, the gravity force per unit volume is a constant vector: $\mathbf{F}_g = \rho\mathbf{g}$.

   The buoyancy term: The buoyancy source term in free convection can be approximated using the relation $\rho_0[1 - \beta(T - T_0)]$. This approximation comes from the linearization of the density variation with respect to temperature based on the definition of the volumetric thermal expansion coefficient, $\Delta\rho = -\beta\rho_0\Delta T$. This is called the Boussinesq approximation. So, the source term due to free convection is

$$\mathbf{F}_g = \rho_0[1 - \beta(T - T_0)]\mathbf{g} \qquad (4.13)$$

where β is the thermal expansion coefficient, and the subscript "0" refers to a reference known condition. This approximation adds a source term to the constant-density (incompressible) form of the Navier–Stokes equation [36]. The energy equation should also be solved in parallel to the Navier–Stokes equation to compute the temperature field.

In order to lay aside the Boussinesq approximation to more precisely simulate the flow field in such applications, we have to consider the compressible form of the Navier–Stokes equation without the mentioned virtual source term. Instead, the original form of the gravity term $\rho\mathbf{g}$ should be used in the equation, and the density is one of the unknowns. This way, we can compute the effect of density variation due to temperature gradient in natural convection without linearization. In this case, the λ-term in the stress–strain rate constitutive relation should be kept alive due to compressibility effects [37].

7. **The electro–magneto–hydrodynamic body force:** The Lorentz force is exerted when a charged particle moves in an electric or magnetic field,

$$\mathbf{F}_{Lo} = \mathbf{J} \times \mathbf{B} + \rho_c \mathbf{E} \tag{4.14}$$

where ρ_c, $\mathbf{J}$, $\mathbf{E}$, $\mathbf{B}$ are the charge density, the electric current density, the electric field, and the divergence-free magnetic field, respectively. Also, we have Maxwell's equations

$$\nabla \times \mathbf{B} = \mu_0 \mathbf{J}, \quad \nabla \times \mathbf{E} = -\frac{\partial \mathbf{B}}{\partial t} \tag{4.15}$$

where μ_0 is the magnetic permeability. The first and the second equations are Ampere's law and Maxwell–Faraday's law, respectively. For more details, see Section 6.2.

8. **The electroosmotic force:** This weak force describes the electro-kinetic motion of an ionized liquid induced by an applied potential in small-scale flows through membranes or in microfluidic devices. The opposite process is called the streaming potential. The origin of such motion is the Coulomb force exerted from the electric field on moving electric charges and ions,

$$\mathbf{F}_{EO} = \rho_c \mathbf{E} \tag{4.16}$$

The potential in the electrical double layer can be obtained using the Poisson–Boltzmann equation

$$\nabla^2 \psi = -\frac{\rho_c}{\varepsilon} \tag{4.17}$$

where ε is the dielectric constant of the solution, and ρ_c is electric charge density and is a function of Faraday's constant and concentration of ions in the base solution. The electrical double layer, or the Debye layer, appears near fluid-solid interfaces where the charge is separated by the ions. Using the Debye–Huckel approximation, the equation for ψ can be linearized to a Helmholtz-type equation as $\nabla^2 \psi = k^2 \psi$ where k is the multiplicative inverse of the Debye length.

9. **The line-loading force:** The surface tension force arises along the free-surface or the interface of two fluids. The surface tension source term or the capillary force could be added to the Navier–Stokes equation as follows,

$$\mathbf{F}_{ST} = \sigma \kappa \nabla H = \sigma \kappa \delta(n) \mathbf{n} \tag{4.18}$$

where κ is the interface curvature being equal to the negative of the divergence of the normal-to-interface unit vector n_i,

$$\kappa = -\frac{\partial n_i}{\partial x_i} \tag{4.19}$$

H is the Heaviside function, δ is the Dirac Delta function, and σ is the surface tension coefficient,

For the interface between two fluids with a large density difference ($\rho_0 > \rho_1$), Brackbill et al. (1992) [38] suggested

$$\mathbf{F}_{ST} = \sigma \kappa \frac{\nabla \rho}{\rho_0 - \rho_1} \frac{2\rho}{\rho_0 + \rho_1} \qquad (4.20)$$

The surface tension coefficient is a function of temperature and pressure. A set of data for variation of the surface tension coefficient of water and methane with respect to pressure and temperature is presented by Sachs et al. [39]. Another methodology in simulating the surface tension force in liquid-gas flows, such as cavitation flows, is presented with the aid of the Korteweg tensor. This way, the surface tension source term is defined as

$$\begin{aligned}
\mathbf{F}_{ST} &= \nabla \cdot \mathbf{K} \\
\mathbf{K} &= \lambda \left[\left(\rho \nabla^2 \rho + \frac{1}{2}(\nabla\rho)^2 \right) \mathbf{I} - \nabla\rho\nabla\rho^T \right]
\end{aligned} \qquad (4.21)$$

where $\mathbf{K}$ is the Korteweg tensor. λ is the capillary coefficient that is related to the surface tension coefficient

$$\lambda = \frac{3}{2} \frac{\sigma h}{\Delta\rho^2} \qquad (4.22)$$

where h is the artificial thickness of the interface, and $\Delta\rho$ is the density difference between the liquid and the vapor [40].

10. **The vorticity confinement term:** Apart from modern and somehow exact models such as the large-eddy simulation, the direct numerical simulation, and the mean flow equations, Steinhoff (1994) [41] presented an idea that the fake diffusion of vorticity can be modeled simply by adding a nonlinear source term called the vorticity confinement term to the Navier–Stokes equation,

$$\mathbf{F}_T = \gamma\rho\mathbf{f} \qquad (4.23)$$

where $\mathbf{f} = -\hat{\mathbf{n}} \times \boldsymbol{\omega}$ is the paddle wheel force, $\hat{\mathbf{n}} = \frac{\nabla\eta}{|\nabla\eta|}$ is the dimensionless vector of vorticity location, $\eta = -|\boldsymbol{\omega}|$ is a scalar field describing the power of vorticity wheel, and γ is a positive tuning parameter. This term is zero inside irrotational regions and confines the outward diffusion of vorticity by adding inward convection of vorticity. If the outward diffusion of vorticity and the added inward convection balance, a covon appears that moves with the fluid stream without spreading.

Due to the definition of the gradient vector, the direction of $\hat{\mathbf{n}}$ is from the low-vorticity point to the high-vorticity point. So, $\hat{\mathbf{n}}$ equals the unit vector pointing away from the center of the vortex. Steinhoff's model can be modified in numerical simulations by using the size of the numerical grid as a correction factor. This way, the source term becomes zero when the size of the grid vanishes. This modified form may be used in numerical methods to avoid using too small computational grids and also to recover the numerically dissipated turbulent structures.

11. **The force-coupling force:** If the flow contains particles, a source term may be added to the Navier–Stokes equation called the force coupling term $\mathbf{F}_{FC}$. The details of the FCM method will be discussed in Section 6.8.2.

12. **Thermophoresis** describes the response of moving particles to the temperature gradient. Brenner et al. (2005) [42] suggested a modified velocity denoted by the volume velocities u^V, which is related to the velocity vector by the following definition (Q4.4),

$$u_i^V = u_i + A_i \tag{4.24}$$

where

$$\mathbf{A} = \alpha \nabla \ln \rho = -kC\nabla T \tag{4.25}$$

$\alpha = \frac{k}{\rho c_p}$ is the thermal diffusivity and generally is not a constant due to the dependency of density to temperature $\rho(T)$. $C = \frac{1}{c_p} \frac{\partial v}{\partial T}|_p$ is a constant. The divergence of the volume velocity is non-zero and equals the temporal rate of volume production per unit volume of the fluid. Brenner and Bielenberg [43] suggested that the source term $\mathbf{F}_{TP} = \frac{\mu}{3} \nabla \nabla \cdot \mathbf{u}^V$ could be added to the Navier–Stokes equation to take into account the thermophoresis effects where $\mathbf{u}^V$ is the volume velocity. Sonophoresis and iontophoresis are other tools to strengthen the motion of particles and augment the diffusion of a compound into a bed using phenomena such as cavitation or ultrasound emissions.

13. **Stochastic** or random form of the Navier–Stokes equation can be constructed by adding a white-noise-type source term to the equation. This form of the equation will be discussed in Section 6.10.2.

4.1.2 ALTERNATIVE FORMS (Q4.5)

4.1.2.1 The Weak Form

A strong solution of a differential equation satisfies the equation at any point inside the domain. The term "strong" is just added to emphasize this characteristic in comparison to the weak solutions. So, the strong solution of an equation simply refers to the solution. If continuous functions like M and N are multiplied by both sides of the equation, the strong solution of the equation also satisfies the new version of the equation. Consider the unsteady incompressible form of the Navier–Stokes equation and the conservation of mass equation,

$$\nabla \cdot \mathbf{u} = 0$$
$$\rho \frac{D\mathbf{u}}{Dt} + \nabla p - \mu \nabla^2 \mathbf{u} = 0 \tag{4.26}$$

In order to construct a weak solution, multiply both sides of equations by the test functions (weight functions) $M(\mathbf{r})$ and $N(\mathbf{r})$, which yields

$$[\nabla \cdot \mathbf{u}] M(\mathbf{r}) = 0$$
$$\left[\rho \frac{D\mathbf{u}}{Dt} + \nabla p - \mu \nabla^2 \mathbf{u} \right] N(\mathbf{r}) = 0 \tag{4.27}$$

Then, integrate equations over the domain of solution Ω,

$$\int_\Omega [\nabla \cdot \mathbf{u}] M(\mathbf{r}) d\Omega = 0$$

$$\int_\Omega \left[\rho \frac{D\mathbf{u}}{Dt} + \nabla p - \mu \nabla^2 \mathbf{u} \right] N(\mathbf{r}) d\Omega = 0 \qquad (4.28)$$

Use the integration by part in higher dimensions or the Green's first identity,

$$\int_\Omega (\nabla N \cdot \nabla \mathbf{u}) d\Omega = \int_\Gamma (N \nabla \mathbf{u} \cdot \mathbf{n}) d\Gamma - \int_\Omega N \nabla^2 \mathbf{u} d\Omega \qquad (4.29)$$

By assuming zero weight functions on boundaries of the domain (Γ), the first term in the right-hand side vanishes.

$$\int_\Omega \nabla^2 \mathbf{u} N(\mathbf{r}) d\Omega = - \int_\Omega \nabla \mathbf{u} \cdot \nabla N(\mathbf{r}) d\Omega$$

$$(4.30)$$

By replacing the high-order Laplacian term, the final weak form of the Navier–Stokes equation is

$$\int_\Omega [\nabla \cdot \mathbf{u}] M(\mathbf{r}) d\Omega = 0$$

$$\int_\Omega \left[\rho \frac{D\mathbf{u}}{Dt} + \nabla p + \mu \nabla \mathbf{u} \cdot \nabla N \right] N(\mathbf{r}) d\Omega = 0 \qquad (4.31)$$

To clarify the vector operator $\nabla \mathbf{u} \cdot \nabla N$, which is the dot product of a second-order tensor and a vector, the result in two-dimensional cartesian coordinates is expanded.

$$\nabla \mathbf{u} \cdot \nabla N = \left(\frac{\partial N}{\partial x} \frac{\partial u}{\partial x} + \frac{\partial N}{\partial y} \frac{\partial u}{\partial y}, \quad \frac{\partial N}{\partial x} \frac{\partial v}{\partial x} + \frac{\partial N}{\partial y} \frac{\partial v}{\partial y} \right)^T \qquad (4.32)$$

The great advantage of this low-order weak form is that the velocity components only need to be differentiable and continuous up to first-order. The weak form of the equation is useful when we cannot obtain a strong solution for any reason such as the mathematical complexity of the equation. So, we employ some mathematical relations to estimate the behavior of the strong solution of the flow field that are not necessarily twice differentiable. It means that a solution of the weak type may not satisfy the initial form of the equation. Thus, the weak solution is not strong!

The weak form of the equation may have multiple solutions, and we need some additional selection criteria to deal with the non-uniqueness of the solution, called the entropy condition. One of the most common applications of the weak form of the Navier–Stokes equation is in numerical methods, such as finite element technique. Polynomials are appropriate candidates for the weight functions M and N, since they are always continuous over any interval up to any order.

Variational principle: There are attempts to present an energy functional for the Navier–Stokes equation. In this strategy, we minimize an energy functional of the equation instead of directly solving the equation itself. This technique belongs to the framework of the variational calculus, which is more common in solid mechanics. Finlayson [44] indicated that the variational solution of the steady Navier–Stokes equation exists if $\mathbf{u} \times (\nabla \times \mathbf{u}) = \mathbf{0}$ or $\mathbf{u} \cdot \nabla \mathbf{u} = \mathbf{0}$. Yasue [45] discussed the variational principle for incompressible viscous flow under a certain class of stochastic variations.

4.1.2.2 Poisson Equation for Pressure

One of the most challenging issues in solving the incompressible form of the Navier–Stokes equations is the one-way coupling of the continuity and momentum equations. Simply speaking, the pressure only exists in the momentum equation and the continuity equation is written based on the velocity vector.

This point is not problematic in the compressible form due to the existence of the equation of state, which relates the pressure in the momentum equation to the density in the continuity equation. So, it is interesting to construct a new relation for incompressible flows to play a role similar to the equation of state in compressible flows. The one-way coupling of equations may become critical in the numerical solution of the Navier–Stokes equation in computational fluid dynamics.

Pressure is a thermodynamic and mechanical property in compressible and incompressible flows, respectively. Consider an isothermal isodensity incompressible flow. If the pressure is a function of density and temperature as two independent thermodynamic properties, then the pressure has to be a constant, and the gradient of pressure in the Navier–Stokes equation vanishes. This fact is mathematically impossible since the Navier–Stokes equation and the continuity will be governed only by one unknown (the velocity vector). We have two over-determined equations. Consequently, the pressure in incompressible flows should not be a function of two other thermodynamic properties.

Let us start from the incompressible form of the Navier–Stokes equation and take the divergence of both sides

$$
\begin{aligned}
0 &= \nabla \cdot \left[\rho \frac{D\mathbf{u}}{Dt} + \nabla p - \mu \nabla^2 \mathbf{u} + \mathbf{F} \right] \rightarrow \\
0 &= \rho \left[\frac{\partial}{\partial t} (\nabla \cdot \mathbf{u}) + \nabla \cdot [(\mathbf{u} \cdot \nabla)\mathbf{u}] \right] + \nabla \cdot \nabla p - \mu \nabla^2 (\nabla \cdot \mathbf{u}) - \nabla \cdot \mathbf{F} \rightarrow \\
\nabla^2 p &= -\rho \nabla \cdot [(\mathbf{u} \cdot \nabla)\mathbf{u}] + \nabla \cdot \mathbf{F}
\end{aligned}
\tag{4.33}
$$

So, the equation of state is a Poisson-type equation for incompressible flows. This equation reduces to the Laplace equation for the Stokes flow with negligible advection acceleration and vanishing source terms. The solution of the equation of state as an algebraic equation is simpler compared to the solution of the Poisson-type equation for incompressible flows. The global nature of elliptic equations (here for the velocity field) is the source of the main difference between the Navier–Stokes equation and the vorticity equation or the energy equation. The absence of a pressure-gradient-like term in the heat and vorticity equations is the reason behind the local nature of vorticity and temperature.

Creation of another Poisson-type equation for pressure is the basis for the projection fractional-step numerical scheme in computational fluid dynamics. In this scheme, to prevent solving an elliptic equation for pressure in steady flows, a new term is added to the continuity equation called the artificial compressibility,

$$\frac{\partial p}{\partial t} + c^2 \nabla \cdot \mathbf{u} = 0 \tag{4.34}$$

4.1.2.3 Weakly Compressible Form

In order to prevent dealing with the second-order Poisson equation for pressure, and to directly compute a divergence-free velocity field, we may use the weakly compressible form of the conservation laws. We invite the density to come into our play, but instead, we need an extra condition to close the equations regarding this new guest. When the speed of sound increases and the Mach number becomes small (at least less than 0.1), it is expected that asymptotically incompressible results can be produced.

The first strategy: This is the simplest strategy of the weakly compressible flows, where we use the compressible version of the continuity and the momentum equations. To close the equations, we need a relation between pressure and density called the equation of state. An appropriate relation is Tait's equation, which produces appropriate results for dense materials such as liquids,

$$\frac{p+A}{p_0+A} = \left(\frac{\rho}{\rho_0}\right)^m \tag{4.35}$$

where the subscript "0" refers to a reference condition. Then, the speed of sound is computed as

$$c^2 = \frac{\partial p}{\partial \rho} = m\frac{p_0+A}{\rho_0^m}\rho^{m-1} = \frac{m(p+A)}{\rho} \tag{4.36}$$

and should be at least ten times the maximum speed of the fluid.

In this formulation, pressure is computed from the equation of state instead of the Poisson equation. So, the accuracy of the pressure computation will directly depend on the degree of accuracy of the state equation for each particular case. This is the bottleneck of the weakly compressible forms. In order to solve this problem, a second strategy will be explained in the next section.

The second strategy: However, we try to construct a more general strategy that exactly satisfies the divergence-free condition for the velocity vector [46]. Consider the compressible form of the Navier–Stokes equations along with the equation of state and the energy equation for a calorically perfect gas. After normalization, $\frac{1}{M^2}$ appears near the pressure gradient term that produces a singularity when the Mach number goes to zero in incompressible limit. So, we have to replace $\frac{1}{M^2}\nabla p$ with ∇p_h where p_h is the hydrodynamic pressure. Asymptotic expansion of pressure with respect to the Mach number is

$$p = p_t + M p_a + M^2 p_h + O(M^3) \tag{4.37}$$

where p, p_t, p_a, p_h are the total, the thermodynamic, the acoustic, and the hydrodynamic pressures, respectively. After some mathematical operations for incompressible flows, the hydrodynamic pressure appears and the M^{-2}-singularity is removed.

4.1.2.4 Linear Aeroacoustics

The motion of fluids always is accompanied by the generation of unwanted sounds called noises. One of our challenges is to predict and reduce the overall level of sound generated during motion of a fluid. The acoustic pressure is the multiplier of the first order term in the expansion (Equation 4.37). However, the direct numerical computation of the acoustic pressure is not straightforward due to its small magnitude in comparison to the thermodynamic and hydrodynamic pressures.

In an alternative strategy, we try to convert the compressible form of the Navier–Stokes equations into the wave equation. The remaining terms will be encountered as source terms of the final wave equation. This strategy is called the acoustic analogy, which is constructed based on the similarity of the wave equation with the modified Navier–Stokes equation. This way, we have made a bridge between acoustics and fluid mechanics called aeroacoustics.[1]

Helmholtz's wave equation: The Helmholtz's wave equation is a hyperbolic equation, which is the simplest equation to describe the acoustic scattering,

$$\frac{\partial^2 U}{\partial t^2} = c^2 \nabla^2 U \tag{4.38}$$

where c is the speed of sound. Assuming U to be in the form $ue^{-i\omega t}$

$$\nabla^2 u + k^2 u = 0 \tag{4.39}$$

where $k = \omega/c$ is the wave number.

Lighthill's analogy: Multiplying the compressible form of the continuity equation

$$\frac{\partial \rho}{\partial t} + \nabla \cdot (\rho \mathbf{u}) = 0$$

by the velocity vector and adding to the Navier–Stokes equation [47],

$$\rho \frac{\partial \mathbf{u}}{\partial t} + \rho (\mathbf{u} \cdot \nabla) \mathbf{u} = -\nabla p + \nabla \cdot \boldsymbol{\tau} + \mathbf{F}$$

we obtain the conservation form of the momentum equation

$$\frac{\partial (\rho \mathbf{u})}{\partial t} + \nabla \cdot (\rho \mathbf{u} \otimes \mathbf{u}) = -\nabla p + \nabla \cdot \boldsymbol{\tau} + \mathbf{F} \tag{4.40}$$

For mathematical details, see the solved problem (Equation 3.16). Taking the divergence of Equation (4.40)

$$-\frac{\partial^2 \rho}{\partial t^2} + \nabla \cdot \nabla \cdot (\rho \mathbf{u} \otimes \mathbf{u}) = -\nabla^2 p + \nabla \cdot \nabla \cdot \boldsymbol{\tau} + \nabla \cdot \mathbf{F} \tag{4.41}$$

The first term is replaced using the continuity equation.

The density-based version: In order to construct an equation similar to the wave equation, we need to add and subtract $c_0^2 \nabla^2 \rho$ to the left-hand side. After rearranging terms

$$\frac{\partial^2 \rho}{\partial t^2} - c_0^2 \nabla^2 \rho = \nabla \cdot \left[\nabla \cdot (\rho \mathbf{u} \otimes \mathbf{u}) - \nabla \cdot \boldsymbol{\tau} + \nabla p - c_0^2 \nabla \rho \right] - \nabla \cdot \mathbf{F} \qquad (4.42)$$

c_0 is the speed of sound of the noise-generating medium. Equation (4.42) is the famous Lighthill's analogy. Using the index notation, the equation above can be rewritten as

$$\frac{\partial^2 \rho}{\partial t^2} - c_0^2 \frac{\partial^2 \rho}{\partial x_i^2} = \frac{\partial^2 T_{ij}}{\partial x_i \partial x_j} - \frac{\partial F_i}{\partial x_i}$$

$$T_{ij} = \rho u_i u_j - \tau_{ij} + (p - c_0^2 \rho) \delta_{ij} \qquad (4.43)$$

where T_{ij} denotes the Lighthill's acoustic stress tensor. Based on Equation (4.43), T_{ij} is composed of the advection part and the viscous noise generation. Lighthill implied that the viscous source of sound is negligible. The other source of noise generation on the right-hand side is the body force.

The point here is that we can decompose the pressure or density into two parts $p = p' + p_0$ and $\rho = \rho' + \rho_0$. p' and ρ' are called the acoustic parameters. All the procedures taken to derive Lighthill's equation, even the decomposing process, are exact without applying any approximation or neglecting any term. If one assumes that the primed parts are small in comparison to the 0-state, we may linearize the solution by adding the assumption of weak acoustic waves. The primmed version of the equation is

$$\frac{\partial^2 \rho'}{\partial t^2} - c_0^2 \frac{\partial^2 \rho'}{\partial x_i^2} = \frac{\partial^2 T_{ij}}{\partial x_i \partial x_j} - \frac{\partial F_i}{\partial x_i} \qquad (4.44)$$

The pressure-based version: Another version of Lighthill's equation can be obtained by adding and subtracting $\frac{1}{c_0^2} \frac{\partial^2 p}{\partial t^2}$ to Equation (4.41). So,

$$\frac{1}{c_0^2} \frac{\partial^2 p}{\partial t^2} - \nabla^2 p = \frac{\partial^2}{\partial x_i \partial x_j} (\rho u_i u_j + \tau_{ij}) - \frac{\partial F_i}{\partial x_i} + \frac{\partial^2}{\partial t^2} \left(\frac{p}{c^2} - \rho \right) \qquad (4.45)$$

There is a linear relation between pressure and density for the isentropic process

$$dp = \left(\frac{\partial p}{\partial \rho} \right)_s d\rho + \left(\frac{\partial p}{\partial s} \right)_\rho ds = c^2 d\rho + 0 \qquad (4.46)$$

where the speed of sound is defined as $c^2 = \left(\frac{\partial p}{\partial \rho} \right)_{s_0}$ and the last term is zero due to isentropic nature of the problem and weakness of acoustic disturbances. Using the approximation $p - p_0 = c^2 (\rho - \rho_0)$ containing the reference pressure and density indicated by the 0 subscript, the last term in Equation (4.45) vanishes.

For inviscid constant-density flows with divergenceless body forces, Lighthill's equation reduces to

$$\frac{1}{c_0^2}\frac{\partial^2 p}{\partial t^2} \; - \; \nabla^2 p = \rho_0 \frac{\partial^2 T'_{ij}}{\partial x_i \partial x_j}$$

$$T'_{ij} \;\; = \;\; u_i u_j \qquad\qquad (4.47)$$

The pressure-based Lighthill's equation, including the perturbation of density and pressure, is

$$\frac{1}{c_0^2}\frac{\partial^2 p'}{\partial t^2} \; - \; \nabla^2 p' = \frac{\partial^2}{\partial x_i \partial x_j}\left(\rho u_i u_j + \tau_{ij}\right) - \frac{\partial F_i}{\partial x_i} + \frac{\partial^2}{\partial t^2}\left(\frac{p'}{c^2} - \rho'\right)$$

$$(4.48)$$

The Lighthill's analogy has been extended to consider the effect of stationary and moving walls, which are called the Curle [48] and the Ffowcs Williams–Hawkings analogies [49], respectively. Also, it is reported that the theory of sound production is closely related to the vorticity dynamics referred to as the vortex sound [50]. Other acoustic models include

- **Explosion:** The Clark's equation, a nonlinear third-order equation as a permutation of Helmholtz's equation, describes the sound propagation during the thermal explosion process [51].
- **Nonlinear acoustics:** The van Wijngaarden–Eringen equation, the Kuznetsov equation, the Westervelt equation, and the Kokhlov–Zabolotskaya–Kuznetsov equation to investigate the nonlinear propagation of acoustic pressure waves in applications such as bubbly liquids [52].
- **Hydroacoustics:** acoustic models for the propagation of sound waves in liquids in underwater applications [53].
- **Duct acoustics and cavity acoustics:** study of the internal flow noise generation [54].
- **Vibroacoustics:** propagation of mechanical waves in structures [55].

4.1.2.5 Water Hammer

Water hammer is the propagation of pressure waves in liquids with the speed of sound due to a sudden change of kinetic energy of flow.[2] The governing equations for water hammer are similar to those of compressible flows. The two-phase modeling, the turbulence modeling, and the pipe-fluid interaction should also be considered in the full simulation of the water hammer.

Another difference arises in the equation of state as a relation between pressure and density. A new effective bulk modulus is defined,

$$k_e = \frac{k_f}{1 + \frac{k_f}{E}\psi} \qquad\qquad (4.49)$$

ψ is a dimensionless parameter that equals the following magnitudes for elastic ducts with thin walls,

- 0 for rigid ducts
- $\frac{D}{e}$ for a conduit with frequent expansion joints
- $\frac{D}{e}(1 - 0.5v)$ for a pipe anchored at its upstream end only
- $\frac{D}{e}(1 - v^2)$ for a conduit anchored against longitudinal movement throughout its length,

where v is the Poisson's constant, k_f is the bulk modulus of fluid, E is the Young's modulus of elasticity, e is the thickness of the pipe wall, and D is the diameter of pipe. Other suggestions have been presented for thick-walled ducts based on the external and internal radius of the duct [56]. The relation between pressure and density is

$$d\rho = \frac{\rho}{k_e} dp \tag{4.50}$$

The speed of sound or the speed of pressure waves in liquid is

$$c = \sqrt{\frac{k_e}{\rho}} = \sqrt{\frac{k_f}{1 + \frac{k_f}{E}\psi} \frac{1}{\rho}} \tag{4.51}$$

Chaudhry [56] proved that the one-dimensional continuity and momentum equations for transient hydraulics is

$$\frac{\partial p}{\partial t} + V\frac{\partial p}{\partial x} + \rho c^2 \frac{\partial V}{\partial x} = 0$$
$$\frac{\partial V}{\partial t} + V\frac{\partial V}{\partial x} + \frac{1}{\rho}\frac{\partial p}{\partial x} + g\sin\theta + \frac{fV|V|}{2D} = 0 \tag{4.52}$$

where f is the Darcy friction factor, V is the fluid velocity in the x-direction, c is the velocity of the pressure wave in an elastic conduit filled with a slightly compressible fluid, θ is the angle of the pipe with the horizontal direction, and D is the diameter of pipe. There are other theories, such as the Zhukovsky equation, the method of characteristics, and the rigid column theory, to investigate the influence of water hammer on pipelines.

4.1.2.6 Shallow Water Equation

The shallow water equation (SWE) is an alternative form of the Navier–Stokes equation, which is used in open channel flows with small flow depths. Three steps of derivation of the SWE equations is as follows.

Step#1: the pressure decomposition: If the depth of water in an open channel is small, the shear in the z-direction is negligible, and the pressure can be decomposed into the equilibrium and the perturbed parts, $p = p_0 + p'$. Then, the pressure gradient along the vertical direction can be estimated by the hydrostatic pressure distribution,

$$\nabla p_0 = \rho\mathbf{g} \rightarrow \frac{dp_0}{dz} = -g\rho \rightarrow p_0 = -\rho gz + c \tag{4.53}$$

where $\mathbf{g} = (0,0,-g)$. The sum of the pressure term, and the gravity force are substituted into the incompressible form of the Navier–Stokes equation,

$$\begin{aligned}
-\nabla p + \rho\mathbf{g} &= -\nabla(p_0 + p') + \rho\mathbf{g} = -\nabla p' \rightarrow \\
\rho\frac{D\mathbf{u}}{Dt} &= -\nabla p' + \mu\nabla^2\mathbf{u}
\end{aligned} \tag{4.54}$$

In applications such as offshore mechanics, arctic engineering, and ocean flows, the Coriolis term may become important.[3] If the density changes across the channel due to salinity or temperature variation, a similar decomposition should be applied to the density ($\rho = \rho_0 + \rho'$) and a new term ($-\rho'g\mathbf{k}$) appears in the right-hand side.

Step#2: eliminating pressure: Now, we are ready to omit pressure from the momentum equation. Suppose that the height of the free-surface and the height of the bed of the channel are h_u and h_l, respectively. Due to the adhesion kinematic boundary condition over the free-surface, we have

$$\frac{Dh_u}{Dt} = w_u(x,y,t) \tag{4.55}$$

where w_u is the vertical velocity of the free-surface. A similar condition can be applied to the bottom side of the channel,

$$\frac{Dh_l}{Dt} = w_l(x,y,t) \rightarrow \frac{\partial h_l}{\partial t} + (\mathbf{u}.\nabla)h_l = w_l \rightarrow \mathbf{u}.\nabla h_l = w_l \tag{4.56}$$

The last equality holds if the topology of the bed does not change with the evolution of time. The hydrostatic distribution of pressure for the constant-density case indicates that the pressure distribution along the vertical coordinate z is $p = p_u + \rho g(h_u - z)$. The value of p_u is a function of the atmospheric pressure, the curvature of the liquid surface, and the surface tension coefficient. This fact will be neglected here, so the gradient of p_u is zero. Then,

$$\begin{aligned}
p &= p_u + \rho g(h_u - z) \rightarrow (-\rho g z + c) + p' = p_u + \rho g(h_u - z) \rightarrow \\
p' &= \rho g h_u + p_u - c \rightarrow \nabla p' = \rho g \nabla h_u
\end{aligned} \tag{4.57}$$

After substituting relation (4.57) into the momentum equation, the pressure term can be omitted,

$$\frac{D\mathbf{u}}{Dt} + \mathbf{a}_c = -g\nabla h_u + \nu\nabla^2\mathbf{u} \tag{4.58}$$

Similar to all non-conservation forms of the governing equations, this non-conservation form does not hold for the simulation of the hydraulic jump, which includes high gradients.

Step#3: continuity equation: Now, integrate the continuity equation with respect to the vertical coordinate from h_l to h_u,

$$\nabla \cdot \mathbf{u} = \frac{dw}{dz} + \nabla_{x,y} \cdot \mathbf{u} = 0 \rightarrow (w_u - w_l) + (\nabla_{x,y} \cdot \mathbf{u})(h_u - h_l) = 0 \tag{4.59}$$

Substituting the adhesion conditions on the free-surface and the bed of the channel. Using the definition $h = h_u - h_l$

$$\left(\frac{Dh_u}{Dt} - \frac{Dh_l}{Dt} \right) + (\nabla_{x,y} \cdot \mathbf{u})(h_u - h_l) = 0 \rightarrow \frac{Dh}{Dt} + h\nabla_{x,y} \cdot \mathbf{u} = 0 \qquad (4.60)$$

Finally, the conservation form of the continuity equation reads

$$\frac{\partial h}{\partial t} + \nabla \cdot (h\mathbf{u}) = 0 \qquad (4.61)$$

This way, we have eliminated the one-way coupling of the velocity and pressure by replacing the pressure term with the height term that also exists in the continuity equation.

Other mathematical models for the simulation of the SWE flows and surface waves dynamics are

- **The Saint–Venant equation:** Saint–Venant published a paper in 1871 and presented his famous one-dimensional equation. He ignored the viscous term in the Navier–Stokes equation, and treated the viscous effects as a source term $-\frac{P}{A}\frac{\tau_w}{\rho}$. $\frac{A}{P}$ is the hydraulic radius, τ_w is the wall shear stress exerted on the wetted area, and can be computed using the Chezy or Manning equations. Defining the slope of the bed $S = -\frac{dh_l}{dx}$, and the friction slope of the energy grade line $S_f = \frac{\tau_w}{\rho g R}$, the final form is obtained

$$\frac{\partial u}{\partial t} + u\frac{\partial u}{\partial x} + g\underbrace{\frac{\partial h_u}{\partial x}}_{1} + g(\underbrace{S_f}_{2} - \underbrace{S}_{3}) = 0 \qquad (4.62)$$

  Term number 1 corresponds the pressure gradient term, and terms number 2 and 3 are the friction term and the gravity source term, respectively.
- **The Korteweg–de Veries equation (KdV)** is a nonlinear mathematical model for the simulation of surface waves in the framework of shallow water flows. The Benjamin–Bona–Mahony equation is an extension of the KdV equation for long surface-waves.
- **The general Degasperis–Procesi equation** describes the motion of surface waves in shallow water applications.

4.1.2.7 Other Alternative Forms

- **Vorticity equation:** By taking the curl of the Navier–Stokes equation, the vorticity equation will be derived, Equation (7.7). The details of derivation and the physical interpretation of each term in the equation will be presented in Chapter 7.

- **Vorticiy-stream function formulation:** Based on the definition of the stream function for 2D incompressible flows

$$u = \frac{\partial \psi}{\partial y},$$

$$v = -\frac{\partial \psi}{\partial x} \tag{4.63}$$

and the two-dimensional definition of vorticity, we have

$$\omega_z = \left(\frac{\partial v}{\partial x} - \frac{\partial u}{\partial y} \right) = -\nabla^2 \psi \tag{4.64}$$

The pressure can be eliminated by taking the cross derivative of the two-dimensional form of the equations. Then, the Navier–Stokes equation reduces to the stream function-vorticity version,

$$\frac{\partial}{\partial t} \nabla^2 \psi + \frac{\partial \psi}{\partial y} \frac{\partial}{\partial x} \nabla^2 \psi - \frac{\partial \psi}{\partial x} \frac{\partial}{\partial y} \nabla^2 \psi = \nu \nabla^4 \psi \tag{4.65}$$

The main advantage of the vorticity-stream function form is the elimination of pressure. However, it is more challenging to impose boundary conditions based on the magnitude of stream function and vorticity on boundaries.

- **The mechanical energy equation:** The dot product of the velocity vector and the Navier–Stokes equation yields the mechanical energy equation. This equation governs a part of the total energy of fluid particles, which is related to their bulk motion. The other part is proportional to the kinetic energy of molecules governed by the thermal energy equation. The mechanical energy equation will be more deeply discussed in Chapter 5.
- **Madelung's equation:** The Madelung equations is an alternative form of the Schrödinger equation as the fundamental equation of quantum mechanics, written based on the hydrodynamic variables. The Schrödinger Equation will be discussed in Section 6.9.3.
- **Micro/nanoscale version:** If the Knudsen number grows, the non-continuum effects gradually appear. Alternative forms of the Navier–Stokes equation may be derived by applying some corrections. Readers may find more details in Section 6.9.1.
- **Linearized perturbed version:** One of the challenging topics in fluid mechanics is the instability of laminar flows. This theory questions the tolerance of laminar flows to bear weak disturbances. In this context, we consider a perturbed solution by adding a disturbance to a base solution of the Navier–Stokes equation (U). In the linearized version, the disturbances are supposed to be small, and hence the high-order terms can be neglected. After some elaborated algebra, the celebrated Orr–Sommerfeld equation as an eigenvalue problem is obtained for two-dimensional parallel flows [13],

$$(U - c)(v'' - \alpha^2 v) - U'' v + \frac{i\nu}{\alpha}(v'''' - 2\alpha^2 v'' + \alpha^4 v) = 0 \tag{4.66}$$

where c is the propagation speed of Tollmien–Schlichting waves, α is the wave number or the spatial frequency of disturbances, and v is the amplitude of instabilities. In eigenvalue problems, the solution itself does not matter, but the condition (eigenvalues) under which the solution behaves in a certain way (the amplitude decades or grows in space or time)[4] is under study.

The Kuramoto–Sivashinsky equation is another mathematical model to investigate the chaotic nature of the flame front instability. The fourth-order nonlinear Kuramoto–Sivashinsky equation reads

$$\frac{\partial u}{\partial t} + \nabla^4 u + \nabla^2 u + \frac{1}{2}|\nabla u|^2 = 0 \tag{4.67}$$

4.1.3   LIMITING CASES (Q4.6)

The limiting cases of the Navier–Stokes equation include

1. **Hydrostatic fluid:** Neglecting the viscous stress tensor, the body forces except for the gravity force and the acceleration,

$$0 = \rho \mathbf{g} - \nabla p \tag{4.68}$$

2. **Solid body rotation/motion:** Assuming the acceleration to be a constant, the solid/rigid body rotation/motion is introduced,

$$\rho(\mathbf{a} + \mathbf{a}_c) = \rho \mathbf{g} - \nabla p \tag{4.69}$$

In the case of solid body rotation, the centripetal acceleration ($\mathbf{a}_c$) should be added, and a purely rotating forced vortex appears.

In order to keep the forced vortex alive, we need to provide an external power source. Otherwise, the vortex decays and becomes time-dependent, and the time-derivative term should also be added to the equation. Near the stationary bottom wall, the velocity is zero due to the no-slip condition, and an inward radial secondary flow appears from the high-pressure zone toward the low-pressure region near the centerline.

3. **Creeping flows:** Adding the viscous term and the body force term to the equation, ignoring the relative accelerations and the material derivative, then, you will obtain the steady form of the Stokes equation for creeping flows

$$0 = \rho \mathbf{g} - \nabla p + \nabla \cdot \boldsymbol{\tau} + \mathbf{F} \tag{4.70}$$

The fundamental solution or Green's function of the linearized Stokes flow equation is called the Stokeslet, which is the response of the creeping flow to a singular point force in the form of the Dirac delta function.

The Stokes flow for external flows: Taking the curl of the Stokes equation, we obtain the biharmonic equation for the stream function in plane creeping flows,

$$\nabla^4 \psi = 0 \tag{4.71}$$

You may prove that the governing equation for the pressure and vorticity is the Laplace equation.

The Reynolds lubrication equation for internal flows: The inertialess Stokes equation is the governing equation for the traditional lubrication theory of Reynolds for the flow of fluids in narrow gaps. Substitute the velocity profile as a superposition of the linear Couette profile and the parabola of the Poiseuille flow in the continuity equation, and neglect the pressure variation in cross-flow direction, the unsteady three-dimensional, density-varying Reynolds equation for the Lubrication problems reads [58]

$$
\frac{\partial}{\partial x}\left(\frac{\rho h^3}{12\mu}\frac{\partial p}{\partial x}\right) \;+\; \frac{\partial}{\partial z}\left(\frac{\rho h^3}{12\mu}\frac{\partial p}{\partial z}\right)
$$

$$
= \frac{\partial}{\partial x}\left(\frac{\rho h(U_d + U_u)}{2}\right) - \rho U_u \frac{\partial h}{\partial x}
$$

$$
+ \frac{\partial}{\partial z}\left(\frac{\rho h(W_d + W_u)}{2}\right) - \rho W_u \frac{\partial h}{\partial z}
$$

$$
+ \frac{\partial(\rho h)}{\partial t} \tag{4.72}
$$

in which $h(x,z,t)$ is the varying width of the gap, y is the transverse direction, x and z are the steamwise and spanwise directions, the subscripts u and d denote the top and the bottom walls respectively, $\frac{\partial(\rho h)}{\partial t} = h\frac{\partial \rho}{\partial t} + \rho(V_u - V_d)$, and (U,V,W) are the wall velocities. Consider that in creeping internal flows, the Reynolds is not necessarily small.

A modification has been applied to the original version of the equation to take into account the dependency of viscosity on pressure [59] and the effect of hydrophobicity (oleophobicity) of the walls, called the Reynolds–Vinogradova theory [60].

4. **The Oseen equation for creeping flows with inertia:** Add an approximate linearized form of the inertial term to the Stokes equation of creeping flows,

$$
\rho(\mathbf{U}\cdot\nabla)\mathbf{u} = \rho\mathbf{g} - \nabla p + \nabla\cdot\boldsymbol{\tau} + \mathbf{F} \tag{4.73}
$$

where $\mathbf{U}$ is a given divergence-free velocity field, which may be the upstream flow velocity or the speed of motion of a particle such as bacteria or microswimmers in a creeping flow. The fundamental solution or Green's function (impulse response) of the linear Oseen's flow equation is called the Oseenlet, which is the response of flow to a singular point force placed in an Oseen flow.

5. **The Euler equation:** This time, consider the material derivative term and ignore the divergence of the viscous stress tensor. The final equation is called the Euler equation,

$$
\rho\frac{D\mathbf{u}}{Dt} = \rho\mathbf{g} - \nabla p + \mathbf{F} \tag{4.74}
$$

6. **The Prandtl's equations of boundary layer:** After applying the boundary layer approximations at high-Reynold numbers, the diffusion of momentum in the x-direction (parallel to the wall) is ignored. In curvilinear parallel-normal-to-wall coordinates, the Navier–Stokes equation reduces to

$$\frac{\partial u}{\partial t} + u\frac{\partial u}{\partial x} + v\frac{\partial u}{\partial y} = -\frac{1}{\rho}\frac{\partial p}{\partial x} + v\frac{\partial^2 u}{\partial y^2} \tag{4.75}$$

$$\frac{\partial p}{\partial y} = \rho\frac{u^2}{\Re} \tag{4.76}$$

where $\Re$ is the radius of curvature of the wall. For axisymmetric boundary layers, Mangler [61] presented the following equations similar to the cylindrical coordinates

$$\frac{\partial(ru)}{\partial x} + r\frac{\partial v}{\partial y} = 0 \tag{4.77}$$

$$\frac{\partial u}{\partial t} + u\frac{\partial u}{\partial x} + v\frac{\partial u}{\partial y} - \frac{w^2}{r}\frac{dr}{dx} = -\frac{1}{\rho}\frac{\partial p}{\partial x} + v\frac{\partial^2 u}{\partial y^2} \tag{4.78}$$

$$\frac{\partial u_\theta}{\partial t} + u\frac{\partial u_\theta}{\partial x} + v\frac{\partial u_\theta}{\partial y} + \frac{uu_\theta}{r}\frac{dr}{dx} = v\frac{\partial^2 u_\theta}{\partial y^2} \tag{4.79}$$

where r, x, y are the known distance of the wall from the center of the revolution, the tangent-to-wall direction, and the normal-to-wall coordinate. u, v, u_θ are the velocity components in x, y, θ directions, respectively.

7. **The one-dimensional case without the pressure term: The Burgers' equation** is obtained by eliminating the pressure term from the one-dimensional Navier–Stokes equation to avoid complexities originating from the pressure–velocity one-way coupling and higher dimensions. Consequently, the continuity equation should also be ignored. This way, the global behavior of the velocity in the Navier–Stokes equation due to the elliptic nature of the Poisson equation for the pressure disappears. The existence of analytical results based on transformations like the Hopf–Cole transformation is another positive point of the Burgers' equation. The Burgers' equation usually is considered in the one-dimensional version without the body force in the following forms.

   a. The viscose nonlinear Burgers' equation: The case with small values of v corresponding to high-Reynolds numbers appears in direct numerical simulation of turbulence called the Burgulence. Also, one may test the results of different large-eddy simulation models on this equation, and then apply them to the Navier–Stokes equations. The non-conservation and conservation forms of the Burgers' equation, respectively are

$$\frac{\partial u}{\partial t} + u\frac{\partial u}{\partial x} = v\frac{\partial^2 u}{\partial x^2}$$

$$\frac{\partial u}{\partial t} + \frac{1}{2}\frac{\partial u^2}{\partial x} = v\frac{\partial^2 u}{\partial x^2} \tag{4.80}$$

b. The stochastic form of the Burgers' equation is constructed by adding a noise term to the right-hand side called the white noise,

$$\frac{\partial u}{\partial t} + f(u)\frac{\partial u}{\partial x} + \lambda\frac{\partial S}{\partial x} = \nu\frac{\partial^2 u}{\partial x^2} \qquad (4.81)$$

More information about stochastic mechanics is presented in Section 6.10.2.

c. The inviscid nonlinear Burgers' equation is a first-order hyperbolic equation in which the viscous term has been ignored. Due to its hyperbolic nature, this form is beneficial to simulate discontinuities such as shock waves in supersonic flows.

$$\frac{\partial u}{\partial t} + u\frac{\partial u}{\partial x} = 0 \qquad (4.82)$$

d. The linearized Burgers' equation: In order to make everything simpler, we can omit the nonlinearity by replacing the advection velocity with a constant, U. The final equation is called the linearized Burgers' equation.

$$\frac{\partial u}{\partial t} + U\frac{\partial u}{\partial x} = \nu\frac{\partial^2 u}{\partial x^2} \qquad (4.83)$$

e. The linear Burgers' equation is obtained by omitting the linear term and is exactly similar to the heat equation.

$$\frac{\partial u}{\partial t} = \nu\frac{\partial^2 u}{\partial x^2} \qquad (4.84)$$

f. The weak form: Using the one-dimensional integration by part, it is easy to show that the weak form of the initial value inviscid Burgers' equation

$$\frac{\partial u}{\partial t} + \frac{\partial}{\partial x}f(u) = 0, \quad u(x,0) = \psi(x) \qquad (4.85)$$

in the range $(-\infty, +\infty)$ with the weight function $w(x)$ being zero at infinity is

$$\int_0^{+\infty}\int_{-\infty}^{+\infty}\left(u\frac{\partial w}{\partial t} + f(u)\frac{\partial w}{\partial x}\right)dxdt + \int_{-\infty}^{+\infty}\psi(x)w(x,0)dx = 0 \quad (4.86)$$

8. **Bernoulli's equation:** Different forms of the Bernoulli's equation are as follows.

a. **The irrotational form:** Consider the incompressible Navier–Stokes equation with constant viscosity that can be written for all points in the flow,

$$\rho\frac{D\mathbf{u}}{Dt} = \rho\mathbf{g} - \nabla p + \mu\nabla^2\mathbf{u} \qquad (4.87)$$

Split the advection term and the Laplacian term by the following vector identities,

$$(\mathbf{u} \cdot \nabla)\mathbf{u} = \nabla\left(\frac{u^2}{2}\right) - \mathbf{u} \times \boldsymbol{\omega}$$

$$\nabla^2\mathbf{u} = \nabla(\nabla \cdot \mathbf{u}) - \nabla \times \boldsymbol{\omega} \tag{4.88}$$

and assume the flow to be irrotational ($\boldsymbol{\omega} = \mathbf{0}$) and incompressible,

$$\rho\frac{\partial\phi}{\partial t} + p + \frac{1}{2}\rho u^2 + \rho g z = c_1(t) \tag{4.89}$$

where $\phi = \nabla\mathbf{u}$ is the potential function.

b. **For ideal flows:** The Navier–Stokes equation for incompressible inviscid (ideal) flow that may be rotational is written as

$$\rho\frac{D\mathbf{u}}{Dt} = -\rho g\mathbf{k} - \nabla p + 0 \tag{4.90}$$

Write down the advection term along a streamline and simplify it using the fact that the velocity component normal to the streamline is zero. Thus,

$$(\mathbf{u} \cdot \nabla)\mathbf{u} = u_s\frac{\partial u_s}{\partial s} + u_n\frac{\partial u_s}{\partial n} = \frac{\partial}{\partial s}\frac{u_s^2}{2}$$

Compute the dot product of the equation with the differential element vector along the streamline ($d\mathbf{s}$) and integrate the equation along the streamline to find

$$\int_s \frac{\partial u_s}{\partial s}ds + \frac{p_s}{\rho} + \frac{1}{2}u_s^2 + g z_s = c_2(\psi, t) \tag{4.91}$$

where ψ is the stream function.

c. **The compressible form:** When you are integrating along a streamline in the previous version, a relation such as $\frac{p}{\rho^\gamma} = cte$ is needed to connect pressure with density. For this case, the compressible form of Bernoulli's equation is obtained based on the integral $\int \frac{dp}{\rho} = \frac{\gamma}{\gamma-1}\frac{p}{\rho}$,

$$\int_s \frac{\partial u_s}{\partial s}ds + \frac{\gamma}{\gamma-1}\frac{p_s}{\rho} + \frac{1}{2}u_s^2 + g z_s = c_3(\psi, t) \tag{4.92}$$

d. **In relative motion:** Taking into consideration the centripetal accelera-tion, it can be shown that Bernoulli's equation containing the centrifugal term reads

$$\frac{p_s}{\rho g} + \frac{1}{2}W^2 - \frac{1}{2}U^2 = c_4(\psi, t) \tag{4.93}$$

where W is the relative velocity in the rotating motion of a fluid inside the blades of impeller of a turbomachinery and $U = r\omega$.

e. Simpler forms are

$$\frac{p}{\rho} + \frac{1}{2}u^2 + gz = c_5$$

$$p + \frac{\rho}{2}u^2 + \rho gz = c_6$$

$$\frac{p}{\gamma} + \frac{1}{2g}u^2 + z = c_7$$

The second and the third forms are the pressure-from and the head-form of the equation, respectively.

f. **The modified Bernoulli's equation** includes the head loss term, the input or the output power, and the heat transfer term. This equation originally is a version of the energy equation, but it can be interpreted as an extended form of the Bernoulli's equation.

$$\frac{p_1}{\gamma} + z_1 + \frac{u_1^2}{2g} = \frac{p_2}{\gamma} + z_2 + \frac{u_2^2}{2g} \pm \frac{\dot{W}}{\dot{m}g} + h_l \qquad (4.94)$$

4.2 CONSERVATION OF ANGULAR MOMENTUM

In this section, we will answer an important question about the symmetry of the stress tensor at a point in a continuum. When the stress tensor is symmetric, nine components needed to describe the stress state reduces to six. To obtain a statement about the symmetry of the stress tensor, we use the angular momentum conservation law at a point.

Figure 4.1 presents an element of a material with a differential size. This element tolerates shear and normal stresses on its surfaces. It is rotating about the z-axis with the angular velocity of A_z. Before writing the angular conservation law, pay attention to some crucial points.

1. We write the equation in the xy plane. Similar relations can be simply derived for two other planes.
2. The stress tensor is written for one single point. So, after yielding the final relations, δx and δy will go to zero and the volume, the mass, and the moments of inertia of the element will vanish.
3. Since we are investigating the stress tensor at a point, the non-uniformity or uniformity of density and stresses do not matter. The final result is valid for both compressible and incompressible flows.
4. If the element rotates, its rotation should be the rigid-body rotation. This can be proved by reductio ad absurdum. If the element rotates with an angular distortion, the stress needed to distort a zero-volume cell should go to infinity, which is impossible. A simpler statement is obvious from Newton's law of viscosity ($\tau \approx \mu \frac{\Delta u}{\Delta n}$) where by decreasing the size of the element in the denominator, the shear stress grows.

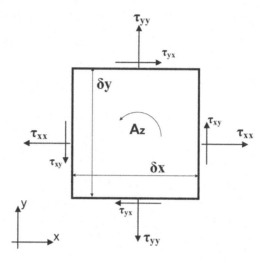

Figure 4.1   Distribution of normal and shear stresses over an element of a continuum.

Start from the conservation of angular momentum or the moment of momentum conservation law,

$$T_z = I_z \frac{dA_z}{dt} \tag{4.95}$$

where T_z, I_z, A_z are the net torque exerted on the element, the moment of inertia about the z-axis, and the angular velocity about the z-axis, respectively. The moment of inertia vanishes as the element shrinks to a point. This fact was obvious from the beginning since a point does not have mass, volume, or moment of inertia.[5]

The moment of the gravity force is zero since the mass of the point-element is zero. Thus, the torque from other body forces such as the electromagnetic forces (e'_z), and the torque from surface forces should be considered in the balance equation.

After multiplying the stresses by their corresponding infinitesimal area and the distance between the couples, we have

$$T_z = \delta x(\tau_{xy}\delta y\delta z) - \delta y(\tau_{yx}\delta x\delta z) + e'_z\delta x\delta y\delta z \tag{4.96}$$

where e'_z is the volumetric external body torque per unit volume, and the moment of the normal stresses is zero. When the size of the element goes to zero ($I_z \to 0$), after dividing both sides of the relation by $d\forall$,

$$T_z = (\tau_{xy} - \tau_{yx} + e'_z)\delta\forall = 0 \to \tau_{yx} = \tau_{xy} + e'_z \tag{4.97}$$

You may write the conservation of angular momentum for two other directions. Using the tensor notation, the final form of the relation is

$$\tau_{ji} = \tau_{ij} + e'_k \tag{4.98}$$

Consequently, the stress tensor is symmetric when the external body torques are zero. Applications in which the stress tensor is not symmetric are magnetic fluids, polar fluids, and ferrofluids under the influence of a magnetic field. A more comprehensive discussion about such applications will be presented in Section 6.5. In addition, a new boundary condition called the no-spin condition should be considered in cases with non-symmetrical stress tensor.

In a more general manner, Łukaszewicz (1999) [57] presented a separate relation for the moment of momentum conservation law

$$\mathbf{e'} + \mathbf{r} \times \left(\rho \frac{D\mathbf{u}}{Dt} - \mathbf{F} - \nabla \cdot \boldsymbol{\tau} \right) = \mathbf{T}_x \tag{4.99}$$

where $\mathbf{T}_x = \varepsilon_{ikj}\tau_{jk}$ and $\mathbf{e'}$ is the external torque vector. $\mathbf{T}_x$ is the dual vector of the stress tensor and represents the anti-symmetric part of the stress tensor. If the two terms in the left-hand side become zero due to the absence of external torques and the conservation of linear momentum, three components $\tau_{32} - \tau_{23}$, $\tau_{13} - \tau_{31}$, $\tau_{21} - \tau_{12}$ of the vector $\mathbf{T}_x$ will vanish. These relations together again imply the symmetry of the stress tensor. Consequently, the moment of momentum equation plus the external torques equal the dual vector of the stress tensor.

We may use the couple stress tensor discussed in Łukaszewicz (1999) [57] to model the behavior of complex fluids. Such fluids, including polyatomic suspensions, anisotropic fluids, liquid crystals, certain types of non-Newtonian fluids, ferromagnetic or ferroelectric fluids, blood, magneto-rheological fluids, and smart fluids are examples of fluids with asymmetric stress tensor.

4.3 BOUNDARY CONDITIONS (Q4.7)

1. **Liquid–solid interface**

   - **Navier's slip condition for liquids:** The slip condition for liquids presented by Navier for an impermeable wall[6] is

$$\begin{aligned} \mathbf{u} \cdot \mathbf{n} &= 0 \\ (\boldsymbol{\tau} \cdot \mathbf{n})_t + k\mathbf{u}_w &= 0 \end{aligned} \tag{4.100}$$

     where k is an adjustable positive coefficient that may be a function of pressure. The subscript "t" denotes the tangential component. The Navier's relations imply that the velocity vector on the boundary is proportional to the tangential component of the stress tensor. The Navier's equation is not a geometric concept and can be derived from the weak formulation of Navier–Stokes equation [62].
     Based on the Navier's slip condition, the slip-length L_s can be defined as an extrapolated distance from the wall where the tangential component of the velocity becomes zero. The slip length follows the relation $u_{slip} - u_{wall} = L_s \frac{\partial u}{\partial n}|_w$. From the mathematical viewpoint, this condition is Robin's third-kind mixed boundary condition. The slip condition can be affected by a combination of surface roughness, wettability, surface energy, and the wall shear rate [63]. Navier's slip condition

may be a candidate as a boundary condition for super-hydrophobic surfaces in contact with liquids [64–66]. Superhydrophobicity or hydrophilicity on the wall can be fabricated by oxidation, coating, machining, and laser pulses.

· **The Stokes' no-slip condition:** The no-slip conditions for liquids presented by Stokes for an impermeable wall read

$$
\begin{aligned}
\mathbf{u} \cdot \mathbf{n} &= 0 \\
(\nabla \times \mathbf{u}) \cdot \mathbf{n} &= 0 \\
\frac{\partial \mathbf{u}}{\partial \mathbf{n}} \cdot \mathbf{n} &= \mathbf{0}
\end{aligned}
\tag{4.101}
$$

where $\mathbf{n}$ is the outward normal unit vector of the boundary. The first condition is equivalent to the impermeability of the wall. The first and the second conditions correspond to zero flux of velocity and vorticity, respectively, across the boundary [62]. Based on the divergence-free condition of vorticity and using the divergence theorem, it can be concluded that the vortex tube cannot be ended at a fixed wall. Since the entering flux of vorticity should be removed from elsewhere. It is possible to prove that the third condition indicates that the normal component of the viscous stress is zero on the wall.

· **The generalized impermeability boundary conditions:** A new boundary condition has been introduced called the generalized impermeability boundary conditions,

$$
\begin{aligned}
\mathbf{u} \cdot \mathbf{n} &= 0 \\
(\nabla \times \mathbf{u}) \cdot \mathbf{n} &= 0 \\
(\nabla \cdot \boldsymbol{\tau}) \cdot \mathbf{n} &= \mathbf{0}
\end{aligned}
\tag{4.102}
$$

The third condition implies that the normal component of the rate of production of the viscous stress on the boundary is zero. The divergence of the viscous stress tensor for an incompressible constant-viscosity fluid is proportional to the Laplacian of the velocity vector. Also, the Laplacian of a divergenceless velocity field equals the negative of the curl of curl of the velocity field. Since the third relation (in Laplacian form) is divergence-free, it is possible to be used for a partial differential equation (the Navier–Stokes equation) with the same order of derivatives. Also, this equation does not appear in the weak form of the Navier–Stokes equation and acts as a natural boundary condition [67]. This type of boundary treatment ($[curl^k\mathbf{u}] \cdot \mathbf{u} = 0$, $k = 0, 1, 2$) is well appropriate for constructing a boundary condition for vorticity in the vorticity equation [68].

2. **Solid–solid interface- Coulomb's slip condition:** There is a similar slip situation between two solids called the Coulomb's slip condition. If the ratio of

the shear stress and the normal stress exceeds a certain value (the Coulomb's friction coefficient), then the two surfaces do not stick and start to slip over each other.

3. **Gas–solid interface- Maxwell's slip condition for gases:** Maxwell used the kinetic theory of gases to obtain the slip condition based on the Navier's slip condition,

$$u_s - u_w \simeq \lambda \frac{\partial u}{\partial n}_w \qquad (4.103)$$

where λ is the mean-free-path of the gas molecules. The slip condition on the wall is related to the adhesive forces on boundaries and has nothing to do with surface roughness. In the dimensionless form, the normalized slip velocity on walls is proportional to the product of the Mach number and Darcy's friction coefficient. So, the no-slip condition in gases is valid when the Reynolds number is high (turbulent flows or far from the leading edge) and the Mach number is small (incompressible, subsonic, and supersonic flows).

4. **Deforming interface of two immiscible fluids, a solid and a fluid, free-surface:** These cases include liquid–gas, liquid–liquid, liquid–vapor, and deforming solid–fluid interfaces. The last case appears in vortex-induced vibration or fluid-structure interaction problems. If the upper fluid is a gas (maybe air in the atmosphere), the interface is called the free-surface. Consider that the interface between two fluids is a two-dimensional non-Euclidean space across which the density has a discontinuity. More details about surface flows is presented in Aris (1989) [69].

- For two points belonging to two connected media located on the interface denoted by 1 and 2 in the xy plane,

$$
\begin{aligned}
u_{t1} &= u_{t2} \\
u_{n1} &= u_{n2} \\
T_1 &= T_2 \\
c_1 &= c_2
\end{aligned}
$$

These relations are the no-slip condition, the impermeability of the interface based on the continuity equation, the no-temperature jump, and the no-concentration jump, respectively. Similar to the slip condition over solid walls, the tangential component of the velocity of two adjacent points are not necessarily the same over the slip-lines, such as what happens in shock-wave interactions in compressible flows, and vortex sheets. On the other hand, if the diffusion of molecules across the interface is not neglected, the normal velocities are not zero, and the deforming surface becomes permeable.

- Conditions for fluxes are derived based on the conservation of momentum, energy, and mass across the interface. It should be noted that these relations are obtained based on zero thickness (volume) of the interface. So, they are exact and will never be violated.

$$\tau_{zx1} = \tau_{zx2}$$
$$\tau_{zy1} = \tau_{zy2}$$
$$q_{z1} = q_{z2}$$
$$\dot{m}_{z1} = \dot{m}_{z2}$$

• The surface tension appears in the balance of normal stresses on the interface,

$$\mathbf{n} \cdot \boldsymbol{\tau}^c \cdot \mathbf{n} = \sigma(\nabla \cdot \mathbf{n}) \qquad (4.104)$$

Neglecting the viscous effects in high-Reynolds number flows and considering the fact that the interface curvature equals the negative of the divergence of the normal-to-interface unit vector, the balance of normal stress on a curved interface yields

$$p = \frac{\sigma}{\Re} \qquad (4.105)$$

where $\Re$ is the radius of curvature of the interface and the right-hand side is called the Laplace pressure. The Laplace pressure is larger for smaller drops, which leads to the diffusion of gas across the interface from high toward low Laplace pressures. This phenomenon is called the Ostwald Ripening. In large-scale problems, the surface tension effect can be neglected, and the equality of pressures of two adjacent points on the free-surface holds.

If the surface tension coefficient varies along the interface due to variation of temperature gradient along the surface, a new phenomenon called the Marangoni convection is induced. The balance of shear stresses at the free-surface with the surface tension gradient yields

$$\mathbf{n} \cdot \boldsymbol{\tau}^c \cdot \mathbf{t} = \nabla \sigma \cdot \mathbf{t} \qquad (4.106)$$

• The kinematic (or the adhesion) condition implies that the particles adhere to the interface and their upward velocity is equal to the interface velocity,

$$w_i(x,y) = \frac{Dz_i}{Dt} = \frac{\partial z_i}{\partial t} + u\frac{\partial z_i}{\partial x} + v\frac{\partial z_i}{\partial y} \qquad (4.107)$$

where the subscript "i" denotes the interface. The advection terms may be neglected in gradually varying free-surface flows in which the free-surface deformation is slight.

5. **Thermal boundary condition- Smoluchowski's jumped temperature:** Smoluchowski presented a condition for temperature on walls similar to Maxwell's relation,

$$T_s - T_w \simeq \frac{2\gamma}{\gamma+1} \frac{\lambda k}{\mu c_p} \frac{\partial T}{\partial n}\bigg|_w \qquad (4.108)$$

Again, it is easy to show that the normalized jumped temperature is proportional to the product of the Mach number and Darcy's friction coefficient. So, the same trends about the validity of the no-slip condition are valid for this case. Another version of the jumped-temperature condition has been presented for nanoscale heat conduction problems in reference [70]. Other options as thermal boundary conditions are specifying heat flux or a combination of iso and iso-flux conditions on boundaries.

6. **Mass transfer conditions:** Similar to the diffusion of heat and momentum, the mass fluxes and concentrations are identical at the interface of two substances. Moreover, in the case of penetration of a dense solid (sublimation) or a dense liquid (evaporation) into an ideal gas phase, the concentration of the penetrating phase can be related to its partial pressure on the interface (Γ) using the Raulti law,

$$p_p|_\Gamma = p_{p,sat} n_p|_\Gamma$$

where n_p is the mole fraction and $p_{p,sat}$ is the saturation pressure at the temperature of the interface.

The reversed process happens when a gas diffuses into a liquid as a host medium. In this scenario, Henry's law can be applied,

$$p_p|_\Gamma = H n_p|_\Gamma$$

where p_p is the partial pressure of the penetrating phase in the surrounding gas, n_p is the mole fraction of the penetrating phase in the liquid, and H is Henry's constant.

If a gas diffuses into a solid, the molar concentration of the gas in the solid at the interface equals

$$\phi_p|_\Gamma = S p_p|_\Gamma$$

where $\phi_p|_\Gamma$ is the molar concentration of the penetrating gas over the interface, and S is the solubility.

7. **Permeable wall:** If suction or blowing through a wall exists, the normal component of the velocity on the wall should be equal to that of the exchanging fluid. If the wall is permeable or porous, hot or cold fluid can be injected to the wall or sucked from the wall. The temperature of the fluid over the wall during suction is equal to the temperature of the exchanging fluid. However, in the case of fluid injection, a balance of energy is needed to determine the temperature of the wall

$$T_w - T_{injected} \simeq \frac{k}{\dot{m}_w c_p} \frac{dT}{dn}\Big|_w \qquad (4.109)$$

where $\dot{m}_w$ is the mass flow rate per unit area of the injected flow on the wall.

8. **The no-spin condition:** This item is meaningful when the stress tensor is not symmetric.

9. **The inlet and outlet conditions** appear when a fluid enters or leaves a domain.
10. **The symmetry/axis conditions:** The symmetry condition is used when a large domain should be simulated, and there is a mirror-like symmetry in the distribution of properties, source terms, boundary conditions, initial conditions, and geometry. Then, you may perform simulations on half the whole domain and save half of the computational time. Over symmetry boundaries, all fluxes (related to the gradients normal to the boundary) and the normal velocity component are zero. The axis condition appears along the axis of revolution in axisymmetric problems. In such cases, a three-dimensional geometry is replaced by a plane slice of the domain, and all circumferential derivatives are zero.

    For example, consider the flow of a fluid with temperature-dependent viscosity in a pipe. Also, there is direct solar radiation at the right angle above the pipe. This flow configuration is not axisymmetric due to the higher/lower fluid viscosity in the warmer region near the top of the pipe. Instead, it is symmetric with respect to the vertical plane. The effect of gravity or natural convection always destroys the top–down symmetry (with respect to the horizontal plane). Also, it can be proved that except dynamic properties such as pressure, in creeping flows, all kinematic properties have a fore-and-aft symmetry.
11. **The periodic condition:** Another trick to just investigate a small portion of a large domain or to omit the need to exert an inlet/outlet condition is to use the periodic boundary condition when a repeating pattern is expected in the domain. Examples are isotropic turbulence modeling without considering the wall effects, simulation of blade cascades in compressors, simulation of flow over one central tube in a bank of tubes, a part of nanofluid flow in a nanochannel, a nanoscale heat sink, and the fully developed Couette flow. In periodic flows, all flow parameters on the inlet and the outlet should be identical. Pay attention that the fully developed Poiseuille flow is not periodic since the pressure linearly decreases, and the pressures at the inlet and the outlet of a periodic cell are unequal.

4.4   SOLUTION METHODS

4.4.1   NUMERICAL TECHNIQUES

We may classify different numerical methods from different aspects.

1. Hybrids methods are a combination of two different methods to keep advantages of both methods in one framework, such as the hybrid molecular dynamics-smoothed-particle hydrodynamics (MD-SPH) method [71].
2. Eulerian methods like the finite-difference method and Lagrangian methods like the SPH method. In Lagrangian methods, the nonlinear term in the advection part of the material derivative operator vanishes.
3. The mesh-based methods like the finite-difference method, the meshfree methods like the SPH method, and the meshless methods like the

boundary-element method (BEM). The mesh-free methods need an initial distribution of particles. There is no connection between these particles, but they can interact via the kernel function as the heart of the numerical integration. However, in meshless methods, an initial distribution of particles is not needed.

4. Continuous methods like the finite-element and discontinuous methods like the distinct-element method.

5. Deterministic method like the finite-volume method and stochastic methods like the Monte–Carlo method.

6. Macroscale methods like the finite-volume technique, microscale methods like the molecular dynamics, and mesoscale methods like the lattice-Boltzmann method.

4.4.1.1 Macroscale Mesh-Based Methods

1. The finite-difference method is based on the replacement of derivatives with finite differences using Taylor's theorem. The main advantages of this technique are simple coding, lower CPU time, and higher accuracy, which make it a good candidate for turbulent flow simulations. However, the finite-difference method is not appropriate for complex geometries with arbitrary curved boundaries, which needs the construction of numerical mappings to make boundaries aligned with the coordinates.

2. The finite-element equations are derived using the weighted integration of the equations with respect to the test/trial/shape functions (perhaps the Lagrange polynomials). The integrated equation is simplified using Green's first identity, which is the high-order version of the integration-by-part. Then, the solution of the weak form of the equation is found by minimizing the residuals using the collocation, the sub-domain, the Galerkin, or the least-square methods. So, in the finite-element method, the physical quantities have been averaged over the elements with respect to the test functions.

Due to the minimization nature of the finite-element method, like what we do in the calculus of variations, the birthplace of this method is structural problems in solid mechanics. The mathematical aspects of this method are more sophisticated in comparison to the finite-volume method. It is easier to extend the scheme to higher degrees of accuracy by choosing higher-order polynomials test functions (the p-refinement). The finite-element method is more flexible in simulating problems with multi-physics and arbitrary geometries. The disadvantage of the finite-element method is unbalanced fluxes over the element faces, which may lead to some discontinuities and instability problems.

3. The finite-volume method is a special case of the finite-element method with unit test functions. In the finite-volume method, the volume integrals of the terms, including the divergence operator, have been converted to the surface integrals of their fluxes using Gauss's theorem. In the finite-volume method, the physical quantities are averaged over the volumes, and the fluxes of mass, momentum, and energy are exactly conserved on surfaces. Similar to the finite-element method, the finite-volume method is appropriate for arbitrary curved

boundaries, and the computational time is lower in comparison to the finite-element method. It is hard to obtain a high-order discretization for the finite-volume method, but the coding process is simpler in comparison to the finite-element method.

4. The spectral methods are the global version of the finite-element method with non-zero trial functions in the whole domain. They use the Fourier transform to map the governing equations into the Fourier domain. This way, the spatial derivatives will be converted to arithmetic operations. The spectral methods have lower computational time and are more accurate. This characteristic makes them a good choice for turbulent flow modeling. However, we need to compute the Fourier transform and the inverse Fourier transform of the unknown function, which may lead to some errors, such as the Gibbs phenomenon, the leakage error, and the aliasing error. The pseudo-spectral methods are the solution to remove the aliasing error from the results.

5. The differential-quadrature method (DQM) is a finite-difference-based method with more flexibility in the selection of nodes and in reaching higher-order approximations. This method is more common in solid mechanics.

6. The immersed-boundary method is designed to simulate the fluid-structure interaction problems using a fixed cartesian grid. In this method, the effect of solid body is added to the momentum equation as a source term. This way a non-body-fitted mesh can capture the motion of boundaries.

7. The concept behind the level-set method is similar to the immersed-boundary method, which employs a cartesian grid to capture the motion of the boundary between two media with variable topology like drops.

4.4.1.2   Macroscale Mesh Free or Meshless Methods

1. The SPH method as a Lagrangian particle-based Navier–Stokes solver approximates unknown parameters such as velocity with a kernel function, including a smoothing length. However, some extra programming is needed in the SPH method to find the neighboring particles, and there are some problems dealing with the boundary conditions [72–75].

2. The vortex method, as a Lagrangian particle-based mesh-free method, traces vorticity particles to construct the solution of the vorticity equation.

3. The boundary-element method as a meshless mesh-reduction method has been designed t solve linear differential equations based on Green's third identity to reduce the dimension of the problem by one. Due to the generation of elements on boundaries, this method is more efficient for geometries with a small surface-to-volume ratio.

4.4.1.3   Microscale Methods

1. The molecular dynamics method is a molecular atomistic Lagrangian method that uses the second law of Newton to trace molecules. The intermolecular forces are modeled using the potential function. Due to high computational

costs of the molecular dynamics method, it is more appropriate for simple geometries.

2. The density functional theory (DFT) is the lowest level ab initio method, which uses the first principle rules of quantum mechanics and the Schrödinger equation to generate the most accurate results in non-continuum media.

4.4.1.4 Mesoscale Methods

1. The lattice Boltzmann method solves the Boltzmann transport equation utilizing particles as clusters of molecules. The lattice Boltzmann method has lower computational time and contains some unknown parameters related to the statistical mechanics.
2. The direct simulation Monte–Carlo method is a stochastic method for solving the Boltzmann equation or other differential equations in applications such as rarified gas dynamics, nanoscale flows, and heat transfer. The Monte–Carlo method is more flexible in dealing with complex geometries in comparison to the molecular dynamics method.
3. The dissipative-particle dynamics (DPD) is an extension of the molecular dynamics method, which omits the potential function and increases the time-step size to reduce the CPU time. In the DPD method, the force term in the second law of Newton is replaced by the sum of three forces: the conservative force, the dissipative force, and the random force. The third term makes the method stochastic.

4.4.2 ANALYTICAL METHODS

There are several mathematical approaches to find an analytical solution for the governing equations. The most well-known analytical methods are listed here.

1. The separation of variables can be used to solve separable partial or ordinary differential equations. In this method, we try to split the unknown function into two functions of just one quantity. Then, we determine the eigenvalues based on boundary conditions.
2. The main idea behind all integral transforms is to map the partial differential equation into a new space where the solution of the equation can be found more simply. There are several integral transforms, such as the Fourier, Laplace, Hankel, Mellin, Hilbert, and Abel transformations.
3. The Fourier integral transform is the general form of the Fourier series for non-periodic functions, which maps the unknown function into the Fourier space as a complex function of the real spatial frequency. This way, the derivatives with respect to the spatial coordinates will reduce to arithmetic operations in the Fourier space. However, at the final stage, we need to compute the inverse Fourier transform of the solution back to the physical domain.

4. The Laplace transform is the general form of the Fourier transform that works with the complex frequency to omit the time-derivative term in transient problems. So, the remaining independent variables of the differential equation in the Laplace domain are just the spatial coordinates. Similar to the Fourier transform, we need to compute the inverse Laplace transform or the Bromwich integral analytically (if possible) or numerically after finding the solution in the Laplace domain [76].

5. The similarity solution tries to find a similarity parameter in the heart of a similarity function to convert the PDE into an ODE. Then, we may solve the ODE using analytical (if possible) or simple numerical methods. Finding the similarity solution is expectable just for the PDEs that their solution obeys a similar pattern like the velocity profile in the boundary layer before the separation point.

6. The Green's function technique finds basic elementary solutions of an inhomogeneous differential equation with Dirac's delta function as a source term. Such basic solutions are called Green's function or the impulse solution. Then, based on the superposition principle, the final solution of the equation is found using the convolution theorem as a convolution of the Green functions and the source term.

7. The perturbation method is appropriate for a type of differential equation containing a term with small order of magnitude in comparison to other terms. This happens in the boundary layer equations when $\varepsilon = \frac{1}{Re}$ is small. In this method, the small term is omitted, and the remaining differential equation is solved to find the unperturbed solution. Then, the solution is expanded using Taylor's theorem around the perturbation parameter ε.

8. The method of characteristics is mainly used for hyperbolic-type equations like the compressible flow equations, Burgers' equation, the wave equation, or Helmholtz's equation. The PDE can be decomposed into two ODEs along its characteristic lines, and the solution can be found by combining the solutions on the characteristic lines [77].

9. The variational solution of a partial differential equation is obtained by minimizing an energy functional of the unknown parameter. A functional of the dependent variable is extracted using the calculus of variations, and the result of minimization is a weak solution of the original PDE.

10. The power series was a common technique for computing the non-similar boundary layer solutions for decades after Prandtl presented his boundary layer theory. The power-series technique supposes that the solution is a power-series with some unknown coefficients: $\sum_{i=0}^{\infty} a_i z^i$.

Most realistic problems in our world are nonlinear. Some other analytical methods appropriate for the nonlinear equations include Charpit's method, Parker–Sochacki's method, Hopf-cole transformation for the nonlinear Burger's equation, and Backlund transformation [78]. Partial differential equations are classified into linear, nonlinear, and quasi-linear equations. In quasi-linear PDEs, coefficients of the nonlinear term are not a function of the highest derivative of the equation. For instance, the coefficient of the nonlinear advection term in the Navier–Stokes equation is not a function of the second-order derivative of the velocity which appears in the viscous term. It means that the equation is linear with respect to the highest-order derivative.

Generally speaking, the sources of nonlinearity in continuum mechanics may be the fractional time derivative, the nonlinear convective acceleration in the inertia term, the nonlinearity of material's constitutive relations such as that of viscoelastic materials and non-Newtonian fluids, nonlinearity of the boundary conditions such as radiation or contact condition, nonlinear body-forces, nonlinear surface tension in adhesion kinematic condition, and the non-local terms in flow inside porous media. Nonlinearities sometimes are linearizable, like the linearized instability analysis, elasticity, creeping motion, acoustics, surface waves (the Airy-wave theory), and small-deflection of beams.

Solutions of the Navier–Stokes equations: Analytic solutions of the Navier–Stokes equation can be classified into elementary and advanced solutions. The elementary solutions have been derived for

- Static fluid,
- The rigid-body linear motion,
- The rigid-body rotation or the forced-vortex, $u_r = 0, u_\theta = cr, \nabla \times \mathbf{u} = 2c\mathbf{e}_z, \boldsymbol{\tau} = \mathbf{0}$. Centripetal acceleration exists.
- The potential vortex or the free-vortex, $u_r = 0, u_\theta = \frac{c}{r}, \nabla \times \mathbf{u} = \mathbf{0}, (\mathbf{u} \cdot \nabla)\mathbf{u} = \mathbf{0}, \boldsymbol{\tau} \neq \mathbf{0}, \nabla \cdot \boldsymbol{\tau} = \mathbf{0}$. The centripetal acceleration exists, and Bernoulli's equation is valid for finding the pressure distribution.
- The sink/source, $u_r = \frac{c}{r}, u_\theta = 0$. All items except the existence of the centripetal acceleration are similar to the previous case.
- The fully-developed Couette flow $u = cy, v = 0, \nabla \times \mathbf{u} = -c\mathbf{e}_z, \tau_{xy} = c\mu, \nabla \cdot \boldsymbol{\tau} = \mathbf{0}, (\mathbf{u} \cdot \nabla)\mathbf{u} = \mathbf{0}$.

More advanced solutions of the Navier–Stokes equation can be classified based on the driving force of the flow,

1. The surface tension-driven flows or the Marangoni flows
2. The pressure-driven flows or the Poiseuille flows
3. The shear-driven flows or the Couette flows
4. The gravity-driven flows
5. The density-driven flows or turbidity currents
6. The Lorentz-driven flows
7. The sound wave-driven flows or acoustic streaming

Different solutions of the Navier–Stokes equation were presented by researchers based on the local derivative (steady versus unsteady flows), the advection acceleration (creeping versus inertial flows), the divergence of the stress tensor or the Laplacian term (viscous versus inviscid flows), suction or injection on walls (impermeable versus porous walls), Newtonian versus non-Newtonian flows, compressible versus density-varying flows, laminar versus turbulent flows, and appearance of slip velocity on walls (microscale versus macroscale flows).

NOTES

1. Hydraulophone is a connection between fluid mechanics and musical arts. Hydraulophone can create sounds with the frequency range, greater than 3 octaves using water jets.
2. A similar phenomenon is seen in hemodynamics called the blood hammer. The collapsing pulse is one of the applications of water hammer in the medical treatment of vascular diseases.
3. Such flows are called the quasi-geostrophic motion, where the Coriolis force, the pressure gradient term, and the inertial term are in balance.
4. These are called the temporal and the spatial instabilities, respectively.
5.

$$I_z = \int_{-\delta x/2}^{\delta x/2} \int_{-\delta y/2}^{\delta y/2} \int_{-\delta z/2}^{\delta z/2} \rho (x'^2 + y'^2) dx' dy' dz' \tag{4.110}$$

The primed parameters are measured with respect to the coordinates attached to the center of the element. By a straightforward multiple integration procedure, the moment of inertia is

$$I_z = \rho \frac{\delta x^2 + \delta y^2}{12} \delta \forall \tag{4.111}$$

which is zero when δx and δy are null. Another equation dealing with the moment of inertia of fluid and its angular velocity is the Kirchhoff–Clebsch equation

6. For non-porous walls, the impermeability condition holds only if the molecular diffusion across the wall is neglected.

5 Heat and Mass Transfer, the Second Law

Similar to the momentum conservation law first, we need to present constitutive relations to relate the heat flux vector or the mass flux vector to temperature or concentration, respectively. The conservation laws for heat and mass transfer applications will be derived using constitutive relations.

5.1 CONSTITUTIVE RELATIONS

The Fourier law and mass diffusion effects: The Fourier law of heat conduction connects the heat flux vector to temperature via a property of matter called thermal conductivity,

$$\mathbf{q}(\vec{r},t) = -\mathbf{k} \cdot \nabla T(\vec{r},t) + \dot{\mathbf{m}}_D C_V T \tag{5.1}$$

where C_V is the specific heat at constant volume and k_{ij} is the thermal conductivity tensor and equals $k\delta_{ij}$ for isotropic materials. The heat flux vector will be substituted in the energy conservation to yield the heat equation. The second term in the right-hand side of Equation (5.1) is added to consider the diffusion of heat due to the Brownian motion of molecules, which is known as the Dufour effect. The Dufour term can be related to the density gradient using Fick's law of diffusion (Q5.1).

Non-Fourier models: Due to the complexities and expensive computational costs of atomistic methods, researchers are trying to develop semi-classical non-Fourier models in lieu of the Fourier-based simulations [81–83]. The well-known non-Fourier methods can be classified as follows.

- **The single-phase-lag model** accounts for the wave behavior of heat that is absent in the Fourier law. An additional phase-lag parameter is added to the argument of the heat flux vector (τ_q),

$$\vec{q}(\vec{r},t + \tau_q) = -k\nabla T(\vec{r},t). \tag{5.2}$$

  After using the energy equation, the resulting equation will be a hyperbolic partial differential equation, sometimes called the Cattaneo–Vernotte (CV) heat equation.
- **The dual-phase-lag model**: As an extension of the CV model, the DPL model adds a second phase-lag parameter to the argument of the temperature gradient (τ_T). This new parameter changes the mathematical nature of the final equation to mixed parabolic–hyperbolic.

$$\vec{q}(\vec{r},t + \tau_q) = -k\nabla T(\vec{r},t + \tau_T). \tag{5.3}$$

The CV and the DPL models overcome the contradiction of the infinite speed of heat carriers (phonons) in the Fourier law. Applications of the

DOI: 10.1201/9781032719405-5

DPL-type constitutive relation can be found in non-equilibrium thermal transport in porous media [84], nanoscale thermal transport [85], biological heat transfer and thermodynamics aspect.

- **The thermon gas model (the thermomass concept)** uses the classical fluid mechanics equations to describe heat transport by defining the equivalent mass of thermal energy called the thermomass based on Einstein's mass-energy relation [86]. When the thermal inertia is negligible, the thermomass equation reduces to the Fourier law. Based on the Debye state equation, the thermomass pressure can be calculated from

$$P_T = \gamma \rho \frac{(CT)^2}{c^2},\tag{5.4}$$

where P_T and γ are the thermomass pressure and Gruneisen constant, ρ, C, c are the density, the specific heat of the structure, and the speed of light, respectively. The thermomass version of the compressible Navier–Stokes equations is

$$\frac{\partial \rho_T}{\partial t} + \nabla.(\rho_T \vec{u}_T) = 0,$$

$$\rho_T \frac{D\vec{u}_T}{Dt} + \nabla P_T + \vec{f}_T = 0.\tag{5.5}$$

where ρ_T, $\vec{u}_T = \frac{\vec{q}}{\rho CT}$, and $\vec{f}_T$ are, respectively, the density of the thermomass, the drift velocity, and the friction force exerted on the thermomass per unit volume.

- **The two-temperature/two-step models**: In nanoscale applications, the temperature of lattice and electron are not identical. Free electrons of a thin layer of material can absorb the incoming energy and transfer it to the lattice. This mechanism creats a relaxation lag between the excitement and the flow of energy. In rapid procedures, the duration of the incoming excitation is shorter than the time-scale of the problem. Hence, the energy transport process should be described in two separate steps. The heat conduction in the material can be described regarding two distinct temperatures that can be unequal under certain conditions.

The two-step heat model is

$$C\frac{\partial T_e}{\partial t} = -\nabla.\vec{q}_e - G(T_e - T_l) + S$$

$$C_l\frac{\partial T_l}{\partial t} = G(T_e - T_l)\tag{5.6}$$

where the subscripts l and e stand for the lattice and the electron, respectively. G and S denote the electron-lattice coupling parameter and the heating source term, respectively. The parabolic and hyperbolic two-step models originate from the Fourier and the CV models, respectively. The hyperbolic version of the two-step model can be obtained if the heat flux vector is substituted from Equation (5.2).

- **The ballistic-diffusive model (BDE):** In the ballistic-diffusive equation (BDE), it is assumed that the internal energy of the heat carriers per unit volume equals the sum of the internal energy of the ballistic ($u_b = CT_b$) and the diffusive ($u_d = CT_d$) components. The constitutive relation for the ballistic and the diffusive internal energies and heat fluxes are

$$\tau \frac{\partial q_d}{\partial t} + q_d = -\frac{k}{C} \nabla u_d$$

$$\tau \frac{\partial u_b}{\partial t} + \nabla \cdot \vec{q}_b = -u_b + \dot{q}_h \qquad (5.7)$$

where $\dot{q}_h$ is the heat generation per unit volume. Again, these constitutive relations should be combined with the energy equation to obtain the final form of the governing equation [87].

- **The Guyer–Krumhansl (GK) equation:** The GK model is an extension of the CV equation to account for the second sound phenomenon and the ballistic propagation in solids. The modified constitutive relation, in this case, is

$$\tau \frac{\partial \vec{q}}{\partial t} + \vec{q} + k \nabla T - \beta_1 \nabla^2 \vec{q} - \beta_2 \nabla \cdot \nabla \vec{q} = 0 \qquad (5.8)$$

where β_1 and β_2 are the Guyer–Krumhansl coefficients [88].

- **Non-local models:** In order to capture the non-local effects in non-continuum media such as nanoscale materials, the heat flux vector in one-dimensional form is written as

$$q''(x,t) = -\int_0^t \int_{-\infty}^{\infty} k^*(x - x', t - t') \nabla T(x', t') dx' dt' \qquad (5.9)$$

where k^* is the effective conductivity. The non-local case can be simplified to the Fourier law at the limit of $k^*(x - x', t - t') = k \delta(x - x') \delta(t - t')$ in which δ is the Dirac delta function and k is the thermal conductivity [89].

- **The non-Fourier fluid flow:** The generalized form of the DPL model for fluid flow with two phase-lags is

$$\tau_q \frac{\delta \mathbf{q}}{\delta t} + \mathbf{q} = -k \left(\nabla T + \tau_T \frac{\delta \nabla T}{\delta t} \right) \qquad (5.10)$$

where the partial time-derivative is replaced by a new operator of the Jaumann or Li-type derivative called the upper-convected material derivative,

$$\frac{\delta}{\delta t} = \frac{\partial}{\partial t} + (\mathbf{u} \cdot \nabla)() - () \cdot \nabla \mathbf{u} + () \nabla \cdot \mathbf{u} \qquad (5.11)$$

Equation (5.11) has analogies with Maxwell's viscoelastic constitutive relation (Christov, 2009) [90]. The following constitutive relation is used in conjunction with the continuity and momentum equations and the energy balance,

$$C\left(\frac{\partial \mathbf{q}^*}{\partial t^*} + \mathbf{u}^* \cdot \nabla \mathbf{q}^* - \mathbf{q}^* \cdot \nabla \mathbf{u}^*\right) \quad + \quad \mathbf{q}^* = -\nabla T^*$$

$$- \quad R\left(\frac{\partial \nabla T^*}{\partial t^*} + \mathbf{u}^* \cdot \nabla\nabla T^* - \nabla T^* \cdot \nabla \mathbf{u}^*\right)$$

(5.12)

where the subscript * denotes the dimensionless form, $C = \frac{\tau_q k}{D^2}$ is the Cattaneo number, and $R = \frac{\tau_T k}{D^2}$ is the dimensionless retardation number. For the case $R = 0$, the equation reduces to the CV model. The Fourier case recovers if $R = C$ (Khayat et al., 2015) [91].

5.2 CONSERVATION OF ENERGY

5.2.1 DERIVATION OF FOURIER-BASED ENERGY EQUATION

Step#1, the equation of total energy (E_t): Starting from the basic form of the first law of thermodynamics

$$dE_{t,M} = \delta Q_M + \delta W_M \quad [kJ] \tag{5.13}$$

The first, the second, and the third terms demonstrate the increase of total energy of the system, the heat added to the system, and the work down on the system. The extensive form of the total energy is defined as $E_{t,M} = M\left(e + \frac{u^2}{2} - \mathbf{g} \cdot \mathbf{x}\right)$.

As usual in flow problems, we need to compute the total energy per unit volume $E_t = \rho\left(e + \frac{u^2}{2} - \mathbf{g} \cdot \mathbf{x}\right)$. Here, e is the internal energy per unit mass, $\frac{u^2}{2}$ is the kinetic energy per unit mass, and $-\mathbf{g} \cdot \mathbf{x}$ is the potential energy per unit mass. The potential energy can be simplified when the gravity acceleration is in -z direction being equal to $(0, 0, -g)$,

$$-\mathbf{g} \cdot \mathbf{x} = -(0, 0, -g) \cdot (x, y, z) = gz \tag{5.14}$$

where $\mathbf{x}$ is the displacement vector of the fluid particle. The sum of potential and kinetic energies is the mechanical energy.

We are usually interested in the rate form of equations. So, we take the time derivative of the equation. Then, we can shift from the Lagrangian approach to the Eulerian viewpoint using the material derivative concept,

$$\frac{DE_{t,M}}{Dt} = \frac{DQ_M}{Dt} + \frac{DW_M}{Dt}, \quad [kW] \tag{5.15}$$

The first term can be expanded,

$$\frac{DE_{t,M}}{Dt} = \frac{DM}{Dt}\left(e + \frac{u^2}{2} - \mathbf{g} \cdot \mathbf{x}\right) + M\frac{D}{Dt}\left(e + \frac{u^2}{2} - \mathbf{g} \cdot \mathbf{x}\right) \tag{5.16}$$

Dividing both sides by volume and substituting from the mass conservation law $\frac{DM}{Dt} = 0$, the time-rate of change of total energy per unit volume reads

$$\frac{DE_t}{Dt} = \rho \frac{D}{Dt}\left(e + \frac{1}{2}u^2 - \mathbf{g}\cdot\mathbf{x}\right) = \rho\left[\frac{De}{Dt} + \frac{D}{Dt}\left(\frac{1}{2}u^2\right) - \mathbf{g}\cdot\frac{D\mathbf{x}}{Dt}\right] \qquad (5.17)$$

So,

$$\frac{DE_t}{Dt} = \rho\left(\frac{De}{Dt} + u\frac{Du}{Dt} - \mathbf{g}\cdot\mathbf{u}\right) \qquad (5.18)$$

The final form of the total energy equation per unit volume is

$$\frac{DE_t}{Dt} = \frac{DQ}{Dt} + \frac{DW}{Dt}, \quad [kW/m^3]$$

$$\rho\left(\frac{De}{Dt} + u\frac{Du}{Dt} - \mathbf{g}\cdot\mathbf{u}\right) = \frac{DQ}{Dt} + \frac{DW}{Dt} \qquad (5.19)$$

Step#2, the heat transfer term: As explained before, the Fourier law is

$$\mathbf{q} = -k\nabla T \qquad (5.20)$$

where $\mathbf{q}$ is the heat rate per unit area. The input rate of heat to the element shown in Figure 5.1a is

$$q_x dydz + q_y dxdz + q_z dxdy \qquad (5.21)$$

and the output rate of heat from the element is

$$\left(q_x + \frac{\partial q_x}{\partial x}dx\right)dydz + \left(q_y + \frac{\partial q_y}{\partial y}dy\right)dxdz + \left(q_z + \frac{\partial q_z}{\partial z}dz\right)dxdy \qquad (5.22)$$

Hence, the net heat transfer rate to the element is

$$\frac{dQ_M}{dt} = -\left(\frac{\partial q_x}{\partial x} + \frac{\partial q_y}{\partial y} + \frac{\partial q_z}{\partial z}\right)dxdydz = -\nabla\cdot\mathbf{q}d\forall \qquad (5.23)$$

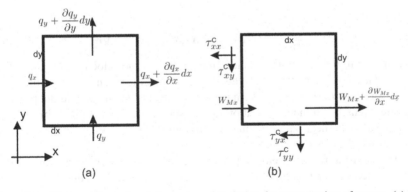

(a) (b)

Figure 5.1 The two-dimensional element used to derive the conservation of energy, (a) the heat flux balance, (b) the stress and work balance.

The negative sign in Equation (5.23) is similar to the negative sign of the pressure force in the Navier–Stokes equation. This negative sign indicates that, for example, if the incoming heat in the x-direction is higher than the output heat rate (the negative heat transfer gradient), the resultant added heat to the system should be positive.

Adding the volumetric source term per unit volume (S) and dividing both sides by the volume of the element, the rate of heat transfer per unit volume becomes

$$\frac{DQ}{Dt} = S - \nabla \cdot \mathbf{q} \tag{5.24}$$

The heat flux vector is eliminated using Fourier's law,

$$\frac{DQ}{Dt} = S + \nabla \cdot (\mathbf{k} \cdot \nabla T) \tag{5.25}$$

Step#3, the work term: The work of the gravity force has already been computed at the first step. The rate of work done on the element by the surface forces with the positive sign is

$$+W_{Mx}dydz + W_{My}dxdz + W_{Mz}dxdy \tag{5.26}$$

The rate of work done by the element with the negative sign is

$$-\left(W_{Mx} + \frac{\partial W_{Mx}}{\partial x}dx\right)dydz - \left(W_{My} + \frac{\partial W_{My}}{\partial y}dy\right)dxdz - \left(W_{Mz} + \frac{\partial W_{Mz}}{\partial z}dz\right)dydx \tag{5.27}$$

The net rate of work per unit volume is

$$\frac{DW}{Dt} = -\nabla \cdot \mathbf{W} \tag{5.28}$$

On the other hand, the net rate of work per unit area on three surfaces of the element shown in Figure 5.1b as a function of stresses is

$$\begin{aligned} W_x &= -u\tau_{xx}^c - v\tau_{xy}^c - w\tau_{xz}^c \\ W_y &= -u\tau_{yx}^c - v\tau_{yy}^c - w\tau_{yz}^c \rightarrow \mathbf{W} = -\mathbf{u} \cdot \boldsymbol{\tau}^c \\ W_z &= -u\tau_{zx}^c - v\tau_{zy}^c - w\tau_{zz}^c \end{aligned} \tag{5.29}$$

Hence,

$$\frac{DW}{Dt} = \nabla \cdot (\mathbf{u} \cdot \boldsymbol{\tau}^c) = \mathbf{u} \cdot (\nabla \cdot \boldsymbol{\tau}^c) + \boldsymbol{\tau}^c : \nabla \mathbf{u} \tag{5.30}$$

The first term in the right-hand side is the product of the velocity vector, and $\nabla \cdot \boldsymbol{\tau}^c$ (the surface force density) represents the mechanical term and changes the kinetic energy of the fluid. The second term in the right-hand side (the double-dot product of the stress tensor and the deformation-rate tensor) is the thermal part. This term will be related to the heat balance equation through compression heating or expansion cooling and the dissipation of heat.

Step#4, final replacement: Replacing the heat transfer term (Equation 5.25) and the work term (Equation 5.30) in the total energy equation (Equation 5.19),

$$\rho\left(\frac{De}{Dt} + u\frac{Du}{Dt} - \mathbf{g} \cdot \mathbf{u}\right) = [S + \nabla \cdot (\mathbf{k} \cdot \nabla T)] + [\mathbf{u} \cdot (\nabla \cdot \boldsymbol{\tau}^c) + \boldsymbol{\tau}^c : \nabla \mathbf{u}] \tag{5.31}$$

5.2.2 ALTERNATIVE FORMS (Q5.2)

Form#1, the thermal energy equation: The dot product of the momentum equation and the velocity vector yields the mechanical energy equation,

$$\rho \frac{D\mathbf{u}}{Dt} = \rho \mathbf{g} + \nabla \cdot \boldsymbol{\tau}^c \rightarrow \mathbf{u} \cdot (\nabla \cdot \boldsymbol{\tau}^c) = \rho \left(u \frac{Du}{Dt} - \mathbf{u} \cdot \mathbf{g} \right) \tag{5.32}$$

The thermal energy equation can be obtained by subtracting the mechanical energy equation from the total energy equation. This way, the first term in the right-hand side of Equation (5.30) will be omitted (the mechanical term). Subtracting the mechanical energy equation (5.32) from the total energy equation (5.31),

$$\rho \left(\frac{De}{Dt} + u \frac{Du}{Dt} - \mathbf{g} \cdot \mathbf{u} \right) = S + \nabla \cdot (\mathbf{k} \cdot \nabla T) + \rho \left(u \frac{Du}{Dt} - \mathbf{g} \cdot \mathbf{u} \right) + \boldsymbol{\tau}^c : \nabla \mathbf{u} \tag{5.33}$$

After canceling similar terms (the mechanical energy) from both sides, we obtain the thermal energy equation,

$$\rho \frac{De}{Dt} = S + \nabla \cdot (\mathbf{k} \cdot \nabla T) + \boldsymbol{\tau}^c : \nabla \mathbf{u} \tag{5.34}$$

The first term is the rate of change of internal energy per unit volume. The second term is the volumetric heat source/sink. The third term represents heat conduction based on Fourier's law, and the last term implies the conversion of kinetic energy to heat.

Split the stress tensor into the hydrostatic (spherical) part and the viscous part,

$$\tau_{ij}^c \frac{\partial u_i}{\partial x_j} = \boldsymbol{\tau}^c : \nabla \mathbf{u} = \boldsymbol{\tau} : \nabla \mathbf{u} - p \nabla \cdot \mathbf{u} \tag{5.35}$$

The first term denotes the irreversible conversion of kinetic energy to heat, and the second term represents the reversible conversion of kinetic energy into internal energy. The last term in Equation (5.35) is related to the pressure and the volume change, which vanishes when $\nabla \cdot \mathbf{u}$ is zero. If the element is expanded $\nabla \cdot \mathbf{u} > 0$, it is negative (expansion cooling). If the element is compressed $\nabla \cdot \mathbf{u} < 0$, it is positive (compression heating). The final form of the thermal energy equation reads

$$\rho \frac{De}{Dt} = S + \nabla \cdot (\mathbf{k} \cdot \nabla T) + \boldsymbol{\tau} : \nabla \mathbf{u} - p \nabla \cdot \mathbf{u} \tag{5.36}$$

Form#2, the enthalpy form: We are going to replace the reversible term using the identity

$$\frac{D}{Dt} \left(\frac{x}{y} \right) = \frac{y \frac{Dx}{Dt} - x \frac{Dy}{Dt}}{y^2} \rightarrow \frac{D}{Dt} \left(\frac{p}{\rho} \right) = \frac{\rho \frac{Dp}{Dt} - p \frac{D\rho}{Dt}}{\rho^2} \tag{5.37}$$

and the continuity equation

$$\nabla \cdot \mathbf{u} + \frac{1}{\rho} \frac{D\rho}{Dt} = 0 \rightarrow \nabla \cdot \mathbf{u} = -\frac{1}{\rho} \frac{D\rho}{Dt}$$

Multiplying both sided by p and replacing $\frac{D\rho}{Dt}$ from identity (5.37)

$$p\nabla \cdot \mathbf{u} = -\frac{p}{\rho}\frac{D\rho}{Dt} = \rho\frac{D}{Dt}\left(\frac{p}{\rho}\right) - \frac{Dp}{Dt} \tag{5.38}$$

Replacing in Equation (5.36),

$$\rho\frac{De}{Dt} = S + \nabla \cdot (\mathbf{k}\cdot\nabla T) + \boldsymbol{\tau}:\nabla\mathbf{u} - \rho\frac{D}{Dt}\left(\frac{p}{\rho}\right) + \frac{Dp}{Dt} \tag{5.39}$$

or

$$\rho\frac{D}{Dt}\left(e + \frac{p}{\rho}\right) = S + \nabla \cdot (\mathbf{k}\cdot\nabla T) + \boldsymbol{\tau}:\nabla\mathbf{u} + \frac{Dp}{Dt} \tag{5.40}$$

The term $\boldsymbol{\tau}:\nabla\mathbf{u}$ is known as the dissipation term Φ and $e + \frac{p}{\rho}$ is enthalpy. Hence,

$$\rho\frac{Dh}{Dt} = S + \nabla \cdot (\mathbf{k}\cdot\nabla T) + \Phi + \frac{Dp}{Dt} \tag{5.41}$$

where the first term is the rate of change of enthalpy per unit volume.

Form#3, incompressible form: The kinetic energy is much smaller than the enthalpy change ($U_c^2 << c_p\Delta T$) in the incompressible limit. This statement is equivalent to a low Eckert number. Using the thermodynamics relation $dh = c_p dT + (1 - \beta T)\frac{dp}{\rho}$, we obtain a familiar quantity: temperature.

$$\frac{Dh}{Dt} = c_p\frac{DT}{Dt} + \frac{1 - \beta T}{\rho}\frac{Dp}{Dt} \rightarrow$$
$$\rho c_p\frac{DT}{Dt} + \frac{Dp}{Dt} - (\beta T)\frac{Dp}{Dt} = S + \nabla \cdot (\mathbf{k}\cdot\nabla T) + \Phi + \frac{Dp}{Dt} \rightarrow$$
$$\rho c_p\frac{DT}{Dt} = (\beta T)\frac{Dp}{Dt} + S + \nabla \cdot (\mathbf{k}\cdot\nabla T) + \Phi \tag{5.42}$$

As an alternative strategy, the temperature can also appear in the internal energy form of the energy equation using another thermodynamic relation between temperature and internal energy. This way, a similar equation may finally be obtained with one important difference: c_p replaced by c_v. On the other hand, the values of the heat capacities are not the same for gases, even in incompressible limit. So, the energy equations derived for gaseous incompressible flows by two different strategies are not identical. Panton (2013) [79] presented a deep discussion of this issue on page 216 and named it the heat equation paradox. The important point is that the correct form should be written based on c_p.

Since the Eckert number is the coefficient of the dissipation and the pressure terms in the dimensionless form of the energy equation (Equation 5.121), both terms can be ignored in the incompressible limit,

$$\rho c_p\frac{DT}{Dt} = S + \nabla \cdot (\mathbf{k}\cdot\nabla T) \tag{5.43}$$

If the thermal conductivity is constant and the material is isotropic (ρ and c_p still may vary with temperature)

$$\rho c_p \frac{DT}{Dt} = S + k\nabla^2 T \tag{5.44}$$

Using the definition of thermal diffusivity $\alpha = \frac{k}{\rho c_p}$, the incompressible energy equation in cartesian coordinates reads

$$\frac{\partial T}{\partial t} + u\frac{\partial T}{\partial x} + v\frac{\partial T}{\partial y} + w\frac{\partial T}{\partial z} = \frac{S}{\rho c_p} + \alpha \left[\frac{\partial^2 T}{\partial x^2} + \frac{\partial^2 T}{\partial y^2} + \frac{\partial^2 T}{\partial z^2}\right] \tag{5.45}$$

Form#4, the e-V form: Adding the mechanical energy equation without the viscous term to the thermal energy equation, replacing the gravity source term by **F**, and using the identity $\nabla \cdot (p\mathbf{u}) = p\nabla \cdot \mathbf{u} + \mathbf{u} \cdot \nabla p$, the e-V form of the energy equation is obtained,

$$\rho \frac{D}{Dt}\left(e + \frac{u^2}{2}\right) = S + \nabla \cdot (\mathbf{k} \cdot \nabla T) + \mathbf{F} \cdot \mathbf{u} - \nabla \cdot (p\mathbf{u}) \tag{5.46}$$

Form#5, the stagnation-enthalpy form: The stagnation enthalpy is related to the static enthalpy as $h_0 = h + u^2/2$. Use the mechanical energy equation to add $\frac{u^2}{2}$ to the enthalpy form and expand the material derivative of pressure. The stagnation-enthalpy form of the energy equation reads

$$\rho \frac{Dh_0}{Dt} = S + \nabla \cdot (\mathbf{k} \cdot \nabla T) + \frac{\partial p}{\partial t} + \mathbf{F} \cdot \mathbf{u} \tag{5.47}$$

Equation (5.47) implies that the sources of generation of stagnation enthalpy are the time-rate of change of pressure in transient problems, the body force, the conduction heat transfer term, and the volumetric generation of heat.

Form#6, the thermodynamic form: We are to omit the velocity vector as a kinematic quantity from the energy equation. This way, the remaining parameters are all thermodynamic quantities. Substitute the velocity divergence from the continuity equation and ignore the dissipation and the conduction terms in the thermal energy equation. Then,

$$\frac{De}{Dt} + p\frac{Dv}{Dt} = \frac{S}{\rho} \tag{5.48}$$

where $v = \frac{1}{\rho}$ is the specific volume. This version is similar to the basic form of the first law of thermodynamics.

Form#7, the rothalpy form: A new parameter is defined in compressible turbo-machineries called the rothalpy,

$$I = h + \frac{u^2}{2} + gz - u_\theta U_{Blade} = h_0 + gz - u_\theta U_{Blade} \tag{5.49}$$

This quantity is conserved when the flow is steady relative to the rotating frame, friction is negligible, and the flow is adiabatic.

5.2.3 REMARKS

- Structure of the incompressible energy equation is similar to the Navier–Stokes equation except the pressure term. This point is the starting point for the development of analogies such as the Reynolds, the Chilton–Colburn, the Prandtl, and the von Karman analogies.
- Similar to the mass conservation law, the energy equation is a scalar equation.
- The obtained energy equation has been derived based on the mass conservation law, the Newtonian-based momentum equation and the Fourier-based energy conservation law. Sometimes all these equations together are called the Navier–Stokes–Fourier (NSF) equations.
- In order to close the energy equation in compressible form, we need the caloric equation of state such as $e = e(p, T)$ or $h = h(p, T)$, $\mu = \mu(p, T)$ as well as the kinetic equation of state $p = \rho R T$.
- In incompressible form, the pressure and the velocity vector are known from the solution of the momentum and the mass conservation laws. The only unknown is temperature. This is the one-way coupling of the mass/momentum equation and the energy equation. However, the coupling is two-way in compressible form. So, ρ, p, T, $\mathbf{u}$ are unknowns, which should be computed by simultaneously solving the mass, the momentum, the energy conservation laws and the equation of state.
- The cylindrical and the spherical forms of heat equation could be easily obtained using the vector operators.
- Similar to the Navier–Stokes equations, the fractional derivatives may be used to produce non-local heat transport behavior.
- If the flow is turbulent, the conductivity should be replaced by the effective conductivity which equals the sum of molecular conductivity and turbulent conductivity.
- The viscous dissipation function is the result of conversion of mechanical energy to heat by viscous effects. It is found that the dissipation term is

$$\Phi = \tau_{ij} \frac{\partial u_i}{\partial x_j} = \boldsymbol{\tau} : \nabla \mathbf{u} = \boldsymbol{\tau} : \boldsymbol{\varepsilon}$$

We can decompose the velocity gradient tensor into the symmetric deformation rate tensor and the anti-symmetric rotation tensor. The double dot product of a symmetric tensor (the stress tensor) and an anti-symmetric tensor (the anti-symmetric part of the velocity gradient tensor) is zero. Then, the velocity gradient tensor can be replaced by the deformation rate tensor. This result is physically expected since the rigid body rotation and motion do not dissipate energy.

The cartesian form of the dissipation function is

$$
\begin{aligned}
\Phi \;=\;& \left(\tau_{xx}\frac{\partial u}{\partial x} + \tau_{yx}\frac{\partial u}{\partial y} + \tau_{zx}\frac{\partial u}{\partial z} \right) + \left(\tau_{xy}\frac{\partial v}{\partial x} + \tau_{yy}\frac{\partial v}{\partial y} + \tau_{zy}\frac{\partial v}{\partial z} \right) \\
+\;& \left(\tau_{xz}\frac{\partial w}{\partial x} + \tau_{yz}\frac{\partial w}{\partial y} + \tau_{zz}\frac{\partial w}{\partial z} \right)
\end{aligned}
\tag{5.50}
$$

Using the symmetry of the stress tensor

$$
\begin{aligned}
\Phi \;=\;& \left(\tau_{xx}\frac{\partial u}{\partial x} + \tau_{yy}\frac{\partial v}{\partial y} + \tau_{zz}\frac{\partial w}{\partial z} \right) + \tau_{yx}\left(\frac{\partial u}{\partial y} + \frac{\partial v}{\partial x} \right) \\
+\;& \tau_{zx}\left(\frac{\partial u}{\partial z} + \frac{\partial w}{\partial x} \right) + \tau_{zy}\left(\frac{\partial v}{\partial z} + \frac{\partial w}{\partial y} \right)
\end{aligned}
\tag{Q5.3}\ \ (5.51)
$$

Replacing the normal and shear viscous stresses from the Stokes constitutive relations

$$
\begin{aligned}
\Phi =\;& \left[\lambda \nabla \cdot \mathbf{u}\frac{\partial u}{\partial x} + 2\mu \left(\frac{\partial u}{\partial x} \right)^{2} \right] + \left[\lambda \nabla \cdot \mathbf{u}\frac{\partial v}{\partial y} + 2\mu \left(\frac{\partial v}{\partial y} \right)^{2} \right] \\
+\;& \left[\lambda \nabla \cdot \mathbf{u}\frac{\partial w}{\partial z} + 2\mu \left(\frac{\partial w}{\partial z} \right)^{2} \right] \\
+\;& \mu \left(\frac{\partial u}{\partial y} + \frac{\partial v}{\partial x} \right)^{2} + \mu \left(\frac{\partial u}{\partial z} + \frac{\partial w}{\partial x} \right)^{2} + \mu \left(\frac{\partial v}{\partial z} + \frac{\partial w}{\partial y} \right)^{2} \\
\Phi =\;& \lambda (\nabla \cdot \mathbf{u})^{2} + 2\mu \left[\left(\frac{\partial u}{\partial x} \right)^{2} + \left(\frac{\partial v}{\partial y} \right)^{2} + \left(\frac{\partial w}{\partial z} \right)^{2} \right] \\
+\;& \mu \left(\frac{\partial u}{\partial y} + \frac{\partial v}{\partial x} \right)^{2} + \mu \left(\frac{\partial u}{\partial z} + \frac{\partial w}{\partial x} \right)^{2} + \mu \left(\frac{\partial v}{\partial z} + \frac{\partial w}{\partial y} \right)^{2}
\end{aligned}
\tag{5.52}
$$

All terms except the first term on the right-hand side of Equation (5.52) are related to the shear viscosity of the fluid and are always positive if $\mu > 0$. This means that the shear stresses produce positive energy loss. The first term originates from normal stresses and may become negative depending on the magnitude of the second coefficient of viscosity. The second law analysis will prove that the dissipation term is a source of entropy generation and is always positive if $\mu \geq 0$ and $\lambda + \frac{2}{3}\mu \geq 0$.

The dissipation term may not be neglected in applications such as fast or small-scale problems, flow of high-Prandtl fluids like oils and polymers, hypersonic flows, wakes, jets, and shear-layers with high velocity-gradients. The dissipation term is the main mechanism of energy dissipation in the energy cascade of turbulent flows.

- The source term may include heat production or absorption during
  - chemical reactions (Section 5.4.2)
  - metabolism energy generation and perfusion in the living tissues (Section 6.7),
  - the Joule heating or thermoelectric effects (Section 6.2),
  - the latent heat of phase change materials ($\rho h_{sf} \frac{\partial f_s}{\partial t}$) in which h_{sf} is the latent heat of melting and f_s is the solid fraction,
  - the radiation heat transfer source term. In order to calculate the radiative heat transfer, an integration over wavelengths from zero to infinity should be performed regarding the absorptivity of the exchanging materials. Some common radiative models are the discrete transfer radiation model, the P-1 model, the surface-to-surface model, the zone method, the differential approximation, the discrete ordinates model, the Rosseland radiation model, and the direct simulation Monte Carlo model.

5.3 THE SECOND LAW OF THERMODYNAMICS

The second law is one of fundamental equations of thermodynamics first developed by Sadi Carno, Rodulf Clasius, and Ludwig Boltzmann. Boltzmann presented a statistical description of entropy. To define entropy, we need the third law of thermodynamics, which implies that the entropy is zero at zero absolute temperature (for pure crystalline solids). Here, we are planning to deal with the continuum nature of entropy. So, we will derive an equation for the time-rate[1] of change of entropy.

The second law of thermodynamics helps us confirm the possibility of different heat/flow arrangements derived based on constitutive relations. For instance, the negative sign in Fourier's law demonstrates that heat transfer takes place from a high-temperature region toward a low-temperature zone. The anti-Fourier behavior happens when the continuum assumption is not valid or the process is far from the equilibrium state.

Step#1, the first law of thermodynamics in Gibbs space: Combining the first law of thermodynamics with the relation $dS = \frac{\delta Q}{T}$,

$$dE = \delta Q + \delta W \rightarrow dE = TdS - pd\forall \tag{5.53}$$

We have neglected other non-$p\forall$ works. In order to take into account some other aspects, such as non-Fourier effects and chemical reactions, we may extend the Gibbs space in the framework of extended irreversible thermodynamics:

$$dE = TdS - pd\forall + \mu dN + ... \tag{5.54}$$

where μ is the molar chemical potential and N is the particle number. Other quantities like the stress tensor and the heat flux vector also may be included in the Gibbs space. The intensive form of the Gibbs relation neglecting kinetic and potential energies (stationary system containing simple compressible substance) reads

$$de = Tds - pdv \tag{5.55}$$

Equation (5.55) is known as the fundamental differential equation of thermodynamics or the first Tds relation. In fluids, the internal energy is usually related to the enthalpy. So, the energy equation for fluids was obtained without the need for the second law of thermodynamics. Acoustic oscillation, sound emission, and absorbtion are exceptions (Chung, 1997) [92]. Based on the definition of enthalpy, $h = e + \frac{p}{\rho}$, the second Tds relation becomes

$$dh = Tds + vdp \tag{5.56}$$

It should be noted that both Tds relations are valid for both reversible and irreversible processes since entropy is a property of matter.

Step#2, derivation of the second law: The material derivative obeys the rule of derivative of the product. So, we replace the specific volume with the density using the following identity,

$$\rho = \frac{1}{v} \rightarrow \frac{Dv}{Dt} = -\frac{1}{\rho^2}\frac{D\rho}{Dt} \tag{5.57}$$

Then, the Eulerian form of the Gibbs relation is

$$T\frac{Ds}{Dt} = \frac{De}{Dt} - \frac{p}{\rho^2}\frac{D\rho}{Dt} \tag{5.58}$$

Multiply both sides by ρ and use the continuity equation ($\rho\nabla\cdot\mathbf{u} + \frac{D\rho}{Dt} = 0$)

$$\rho T\frac{Ds}{Dt} = \rho\frac{De}{Dt} - p\left(\frac{1}{\rho}\frac{D\rho}{Dt}\right) \rightarrow$$

$$\rho T\frac{Ds}{Dt} = \rho\frac{De}{Dt} + p\nabla\cdot\mathbf{u} \tag{5.59}$$

The last term represents entropy production due to the volume change in applications like Joule-expansion. Use the energy equation:

$$\rho\frac{De}{Dt} = -\nabla\cdot\mathbf{q} + \boldsymbol{\tau}:\nabla\mathbf{u} - p\nabla\cdot\mathbf{u} + S$$

to replace the material derivative of the internal energy. Then,

$$\rho T\frac{Ds}{Dt} = -\nabla\cdot\mathbf{q} + \boldsymbol{\tau}:\nabla\mathbf{u} + S \tag{5.60}$$

Divide both sides by T and indicate the viscous dissipation function by Φ

$$\rho\frac{Ds}{Dt} = -\frac{1}{T}\nabla\cdot\mathbf{q} + \frac{\Phi}{T} + \frac{S}{T} \tag{5.61}$$

Using the vector identity

$$\nabla\cdot\left(\frac{\mathbf{q}}{T}\right) = \frac{1}{T}\nabla\cdot\mathbf{q} - \frac{1}{T^2}\mathbf{q}\cdot\nabla T \tag{5.62}$$

Finally, the entropy equation reads

$$\rho \frac{Ds}{Dt} = \underbrace{-\nabla \cdot \left(\frac{\mathbf{q}}{T} \right)}_{1} \underbrace{- \frac{1}{T^2} \mathbf{q} \cdot \nabla T}_{2} + \underbrace{\frac{S}{T}}_{3} + \underbrace{\frac{\Phi}{T}}_{4} \tag{5.63}$$

The entropy field equation presents a relation for the time-rate of change of entropy as a function of entropy flux (positive or negative), and entropy generation (always positive). The mentioned terms in entropy equation are discussed as follows.

1. The sign of this term can change. So, it can enhance or reduce the entropy. It represents the entropy change of material as a reversible effect of heat transfer. It is also related to the entropy change caused by heat transfer that is not necessarily accompanied by temperature gradient. This appears at perfect contact without thermal resistance. Consider an element of a fluid with uniform temperature distribution. In one-dimension, if the gradient of the heat flux component in the positive x-direction is positive, the incoming heat flux is smaller than the output heat flux. So, the net heat flux to the element is negative, and heat is extracted from the element. Referring to the negative sign in the equation, the entropy decreases in this case. The same discussion is valid when the sign of the heat flux vector and its gradients are reversed.

2. The microscopic origin of this term is the phonon–phonon scattering mechanism. It is always positive and represents the increase in entropy due to heat transfer resistance. This term is the origin of appearance of gradient of temperature in matter. It becomes zero if the temperature gradient is absent. Using Fourier's law, this term can be simplified to $\frac{k}{T^2} \nabla T \cdot \nabla T$, which is positive if the thermal conductivity is positive.

3. This term is the reversible reduction or increase of entropy due to the sink or source of energy, respectively.

4. The viscous dissipation shows the irreversible conversion of kinetic energy to heat. This term is always non-negative. The proof of non-negativeness of the dissipation term Equation (5.52) in cartesian coordinates if $\lambda + \frac{2}{3}\mu \geq 0$ is as follows. If λ is a positive number, then the statement is proved. If λ is a negative constant obeying $-\frac{2}{3}\mu \leq \lambda < 0$, we have to consider the smallest negative limit when $\lambda = -\frac{2}{3}\mu$. So, if we prove that the dissipation function is a positive quantity when the lower limit of the Stokes hypothesis holds, then the dissipation term will be positive for all other cases. Hence,

$$\begin{aligned} \Phi &= \lambda \left(\frac{\partial u}{\partial x} + \frac{\partial v}{\partial y} + \frac{\partial w}{\partial z} \right)^2 + 2\mu \left[\left(\frac{\partial u}{\partial x} \right)^2 + \left(\frac{\partial v}{\partial y} \right)^2 + \left(\frac{\partial w}{\partial z} \right)^2 \right] \\ &+ \mu \left(\frac{\partial u}{\partial y} + \frac{\partial v}{\partial x} \right)^2 + \mu \left(\frac{\partial u}{\partial z} + \frac{\partial w}{\partial x} \right)^2 + \mu \left(\frac{\partial v}{\partial z} + \frac{\partial w}{\partial y} \right)^2 \end{aligned} \tag{5.64}$$

The last three terms are non-negative if $\mu \geq 0$. Replace $\lambda = -\frac{2}{3}\mu$ using the Stokes hypothesis

$$\Phi = -\frac{2}{3}\mu\left[\left(\frac{\partial u}{\partial x}\right)^2 + \left(\frac{\partial v}{\partial y}\right)^2 + \left(\frac{\partial w}{\partial z}\right)^2 + 2\frac{\partial u}{\partial x}\frac{\partial v}{\partial y} + 2\frac{\partial u}{\partial x}\frac{\partial w}{\partial z} + 2\frac{\partial v}{\partial y}\frac{\partial w}{\partial z}\right]$$

$$+ 2\mu\left[\left(\frac{\partial u}{\partial x}\right)^2 + \left(\frac{\partial v}{\partial y}\right)^2 + \left(\frac{\partial w}{\partial z}\right)^2\right] + positive\ terms \rightarrow$$

$$\Phi = \frac{2}{3}\mu\left[2\left(\frac{\partial u}{\partial x}\right)^2 + 2\left(\frac{\partial v}{\partial y}\right)^2 + 2\left(\frac{\partial w}{\partial z}\right)^2 - 2\frac{\partial u}{\partial x}\frac{\partial v}{\partial y} - 2\frac{\partial u}{\partial x}\frac{\partial w}{\partial z} - 2\frac{\partial v}{\partial y}\frac{\partial w}{\partial z}\right]$$

$$+ positive\ terms \rightarrow$$

$$\Phi = \frac{2}{3}\mu\left[\left(\frac{\partial u}{\partial x} - \frac{\partial v}{\partial y}\right)^2 + \left(\frac{\partial u}{\partial x} - \frac{\partial w}{\partial z}\right)^2 + \left(\frac{\partial v}{\partial y} - \frac{\partial w}{\partial z}\right)^2\right] + positive\ terms$$

$$(5.65)$$

which is again positive if $\mu \geq 0$.

Remarks

- $\frac{Ds}{Dt} = 0$ holds for isentropic flows. In homentropic flows, the material derivative of entropy equals zero since entropy is constant in space.
- It can be concluded that flows without dissipation, source/sink, and heat conduction have to be isentropic. Other sources of entropy production are electron-hole collisions in thin film semiconductors, the dissipation of electrical energy due to Joule heating, mass diffusion, and reaction.
- Omitting two positive terms and changing $=$ to $\geq$, we obtain the inequality form of the second law,

$$\rho\frac{Ds}{Dt} \geq -\nabla\cdot\left(\frac{\mathbf{q}}{T}\right) + \frac{S}{T} \qquad (5.66)$$

- Temperature appears in the denominator of the two terms in the right-hand side. By increasing temperature, the rate of change of entropy decreases. This proves that energy at higher temperatures has a higher quality.
- The sum of two positive terms in Equation (5.63) is called the entropy generation. Fourier's law can be replaced to yield the cartesian form of the entropy generation term,

$$\dot{s}_{gen} = \frac{\Phi}{T} + \frac{k}{T^2}\left[\left(\frac{\partial T}{\partial x}\right)^2 + \left(\frac{\partial T}{\partial y}\right)^2 + \left(\frac{\partial T}{\partial z}\right)^2\right] \qquad (5.67)$$

5.3.1 ALTERNATIVE FORMS

- **Crocco's equation** is the connection between entropy as a thermodynamic property and vorticity as an important quantity in fluid mechanics. The

Navier–Stokes equation without the viscous term is

$$\rho \frac{\partial \mathbf{u}}{\partial t} + \rho (\mathbf{u} \cdot \nabla)\mathbf{u} = -\nabla p + \mathbf{F}$$

Form#1, the enthalpy form of Crocco's theorem: Replace the pressure term in $\nabla h = T\nabla s + \frac{\nabla p}{\rho}$ using the Navier–Stokes equation,

$$T\nabla s = \nabla h - \frac{1}{\rho}\left[-\rho\frac{\partial \mathbf{u}}{\partial t} - \rho(\mathbf{u}\cdot\nabla)\mathbf{u} + \mathbf{F} \right] \tag{5.68}$$

So,

$$T\nabla s = \nabla h + \frac{D\mathbf{u}}{Dt} - \frac{\mathbf{F}}{\rho} \tag{5.69}$$

Form#2, the stagnation enthalpy form of Crocco's theorem: Defining the total or the stagnation enthalpy as $h_0 = h + \left(\frac{\mathbf{u}\cdot\mathbf{u}}{2}\right)$, then

$$\nabla h = \nabla h_0 - \nabla\left(\frac{u^2}{2}\right) \tag{5.70}$$

Hence,

$$T\nabla s = \left[\nabla h_0 - \nabla\left(\frac{u^2}{2}\right)\right] + \frac{D\mathbf{u}}{Dt} - \frac{\mathbf{F}}{\rho} \tag{5.71}$$

Using the identity $(\mathbf{u}\cdot\nabla)\mathbf{u} = \nabla(\frac{1}{2}\mathbf{u}\cdot\mathbf{u}) - \mathbf{u}\times\boldsymbol{\omega}$,

$$T\nabla s = \nabla h_0 - \mathbf{u}\times(\nabla\times\mathbf{u}) + \frac{\partial \mathbf{u}}{\partial t} - \frac{\mathbf{F}}{\rho} \tag{5.72}$$

Equation (5.72) states that for steady inviscid flows with negligible body forces (true for high-speed gaseous flows with thin boundary layers), the gradients of entropy and total enthalpy are linked to vorticity. This is exactly the case that happens behind a curved shock wave in supersonic compressible flows.

The changing slope of a curved shock wave inversely represents its strength. So, the amount of entropy generation across the shock differs at different points along it. So, a non-zero gradient of entropy and consequently a rotational region appears behind the curved shock wave.

Form#3, the plane steady form of Crocco's theorem: On page 193 of Liepmann and Roshko (2001) [93], it is proved that for plane steady flows without body forces, the Crocccco's theorem along a streamline and normal to it can be written as

$$T\frac{ds}{dt} = 0$$

$$T\frac{ds}{dn} = \frac{dh_0}{dn} + u\omega \tag{5.73}$$

It means that entropy does not change along streamlines in plane steady flows with vanishing body forces.

Form#4: the stagnation pressure form of Crocco's theorem: Neglecting the body forces in steady inviscid flow, and again applying the identity $(\mathbf{u} \cdot \nabla)\mathbf{u} = \nabla(\frac{1}{2}\mathbf{u} \cdot \mathbf{u}) - \mathbf{u} \times \boldsymbol{\omega}$ to the Euler's equation

$$\underbrace{\mathbf{u} \times \boldsymbol{\omega}}_{\text{Lamb Vector}} = \frac{\nabla p_0}{\rho} \tag{5.74}$$

where $p_0 = p + \rho\frac{u^2}{2}$ is the total or the stagnation pressure. The cross product of the velocity vector and the vorticity vector is the Lamb vector. When the Lamb vector is zero, the flow is called Beltrami flow. Based on the current version of Crocco's theorem, if the gradient of total pressure is zero in an inviscid steady flow without body forces, we have a Beltrami flow. Also, the total pressure gradient is a source of vorticity production. It should be noted that the total pressure in compressible flows is another representation of entropy.

· **Internal dissipation:** It is customary to define the internal dissipation function $D = \rho T\frac{Ds}{Dt} - \rho Q + \nabla \cdot \mathbf{q}$, which just contains thermodynamic quantities.
Example: Obtain a relation between the internal dissipation and the Helmholtz free energy. Using the Helmholtz free energy $\Psi = e - Ts$

$$\frac{De}{Dt} = \frac{D\Psi}{Dt} + T\frac{Ds}{Dt} + s\frac{DT}{Dt} \tag{5.75}$$

and the first law of thermodynamics $\rho\frac{De}{Dt} = -\nabla\mathbf{q} + \rho Q + \tau_c : \nabla\mathbf{u}$

$$\begin{aligned} D &= \rho T\frac{Ds}{Dt} - \left[\rho\frac{De}{Dt} - \tau_c : \nabla\mathbf{u}\right] \\ &= \rho T\frac{Ds}{Dt} - \left[\rho\frac{D\Psi}{Dt} + \rho T\frac{Ds}{Dt} + \rho s\frac{DT}{Dt} - \tau_c : \nabla\mathbf{u}\right] \\ &= \tau_c : \nabla\mathbf{u} - \rho\left[\frac{D\Psi}{Dt} + s\frac{DT}{Dt}\right] \end{aligned} \tag{5.76}$$

Example: Prove that the internal dissipation may become negative. Based on the second law of thermodynamics

$$\rho\frac{Ds}{dt} - \frac{\mathbf{q} \cdot \nabla T}{T^2} + \frac{\nabla \cdot \mathbf{q}}{T} - \frac{\rho Q}{T} \geq 0 \rightarrow$$

$$D \geq \frac{\mathbf{q} \cdot \nabla T}{T} \tag{5.77}$$

The last term is always negative. So, the internal dissipation term is greater than or equal to a non-positive value, and thus, may become a negative number.

Example: Obtain the internal dissipation for compressible inviscid flows. Using the continuity equation and the first law of thermodynamics

$$T\frac{Ds}{Dt} = \frac{De}{Dt} - \frac{p}{\rho^2}\frac{D\rho}{Dt} \rightarrow$$

$$\rho T\frac{Ds}{Dt} = \rho\frac{De}{Dt} - \frac{p}{\rho}(-\rho\nabla\cdot\mathbf{u}) \rightarrow$$

$$\rho\frac{De}{Dt} = \rho T\frac{Ds}{Dt} - p\nabla\cdot\mathbf{u} \rightarrow$$

$$\rho T\frac{Ds}{Dt} - p\nabla\cdot\mathbf{u} - \rho\frac{De}{Dt} = 0 \rightarrow$$

$$\rho T\frac{Ds}{Dt} + \nabla\cdot\mathbf{q} - \rho Q = 0 \rightarrow$$

$$D = 0 \tag{5.78}$$

In conclusion, based on the first law of thermodynamics without the viscous stress term, the internal dissipation is zero for compressible inviscid flows.

· **Exergy equation:** The concept of exergy is another statement for quantifying the second law of thermodynamics. The exergy of a system is defined as the maximum available useful work during a thermodynamic process to achieve equilibrium with the environment. The specific exergy is defined as the sum of kinetic, potential, strain, and thermal parts, and the chemical exergy component (ignored in the following equation),

$$X = K + P - (p - p_0)v + (T - T_0)S$$

Using the material derivative, we can compute the total rate of change of exergy in Eulerian form

$$\frac{DX}{Dt} = \frac{DK}{Dt} + \frac{DP}{Dt} - \frac{D[(p - p_0)v]}{Dt} + \frac{D[(T - T_0)S]}{Dt}$$

Each term in this relation can be decomposed into its constructing subterms to form a field equation for exergy in continuum media. Such equation contains the following terms,

1. **Transport of kinetic exergy** including shear stress kinetic exergy, kinetic exergy related to the work of non-conservative body forces, kinetic exergy from potential energy conversion, reversible kinetic exergy from strain exergy conversion, and irreversible kinetic exergy from friction.
2. **Transport of potential exergy** due to diffusion, reversible conversion of kinetic and potential exergies, and reactions.
3. **Thermal exergy** due to heat diffusion, temperature difference, conductive heat transfer, conversion from shear strain, normal strain, and non-conservative potential gradients, body forces, chemical reactions, variation of chemical potential gradient, and different reference temperature.

4. **Strain exergy** due to normal stresses, normal strain exergy conversion, irreversible conversion of strain exergy, and change in reference temperature

Also, other five terms correspond to the transport of chemical exergy. More details about the derivation of each of 27 terms have been presented in reference [94].

- **The extended irreversible thermodynamics:** The second law of thermodynamics seems to be violated in some applications in the context of classical Gibbs space. It means that the rate of production of entropy may locally become negative. However, new constitutive relations like the CV model or the DPL model need the permission of the second law of thermodynamics. It is proved that they are locally consistent with the second law of thermodynamics in the framework of extended irreversible thermodynamics.

The extended irreversible thermodynamics extends the existing variables in the space of state to include some other physical aspects. This procedure helps us to perform a second law analysis for some special problems such as heat transfer and fluid flow with

- small length-scales (the non-local and non-Fourier effects in nanotechnology),
- short time-scales (the memory effects),
- the nonlinear effects such as shock waves.

For materials with phase-lag or the flow of Stokesian fluids, the heat flux vector or even the stress tensor may contribute to entropy [91]

$$ds = \left(\frac{\partial s}{\partial T}\right) dT + \left(\frac{\partial s}{\partial \rho}\right) d\rho + \left(\frac{\partial s}{\partial \mathbf{q}}\right) \cdot d\mathbf{q} + \left(\frac{\partial s}{\partial p^v}\right) dp^v + \left(\frac{\partial s}{\partial \boldsymbol{\tau}'}\right) : d\boldsymbol{\tau}'$$
(5.79)

where $\boldsymbol{\tau}'$ is the viscous stress tensor excluding the term related to the second coefficient of viscosity and $p^v = -\lambda \nabla \cdot \mathbf{u}$ [91, 95].

5.4 MASS TRANSFER

In previous sections, the constitutive relations for the diffusion of momentum and heat were introduced. Here, we are to discuss the bulk motion and molecular diffusion of mass. Some examples of mass transfer are evaporation, chemical absorption by the volume of the substance, precipitation, distillation, cooling towers, fuel cells, permeable membrane filtration, purification of liquids (like blood in our kidneys), mass diffusion during welding, and dense flows.

5.4.1 CONSTITUTIVE RELATION

Adolf Fick presented the well-known law of diffusion after performing his famous experiments on the investigation of diffusion of salt between two reservoirs connected by a pipe with non-identical concentrations of salt. It should be noted that in applications such as the diffusion of paint in water or perfume in the air, the underlying molecular mechanism is the random Brownian motion of particles.

Similar to other constitutive relations, Fick's law can be violated in some special applications such as flow in porous media and non-continuum cases. Materials that do not obey Fick's law are called non-Fickian materials. Similarities in the mathematical structure of the laws, for mass, heat, and momentum transfers are the basis of analogies between them. The Sherwood number of mass transfer ($Sh = \frac{hL}{D}$) is analogous to the Nusselt number in heat transfer. The Schmidt number ($\frac{\nu}{D}$, momentum–mass) and the Lewis number ($\frac{\alpha}{D}$, thermal–mass) are defined as analogous to the Prandtl number ($\frac{\nu}{\alpha}$, thermal-momentum).

Fick's first law: Fick's first law indicates that the flux of mass is proportional to the gradient of concentration,

$$\mathbf{J} = -D\nabla\phi \tag{5.80}$$

where ϕ is the molar concentration ($\frac{mol}{m^3}$), D is the mass diffusivity ($\frac{m^2}{s}$), and $\mathbf{J}$ is the molar diffusion flux with the unit $\frac{mol}{m^2 s}$. The negative sign implies that the mass immigrates from the regions of high concentration toward low concentration. Based on the Stokes–Einstein law the diffusion coefficient is a function of product of temperature and the Boltzmann constant.

Fick's second law: Using the mass conservation law,

$$\frac{\partial\phi}{\partial t} + \nabla\cdot\mathbf{J} = 0 \tag{5.81}$$

and replacing the mass flux with the aid of Fick's first law, we obtain

$$\frac{\partial\phi}{\partial t} - \nabla\cdot(D\nabla\phi) = 0 \tag{5.82}$$

Assuming constant diffusivity, Fick's second law appears that is a partial differential equation for concentration distribution,

$$\frac{\partial\phi}{\partial t} = D\nabla^2\phi \tag{5.83}$$

Anisotropic material: For anisotropic material, the diffusivity is a second-order tensor, and Fick's first and second laws are

$$\mathbf{J} = -\mathbf{D}\cdot\nabla\phi \quad or \quad J_i = -D_{ij}\frac{\partial\phi}{\partial x_i}$$
$$\frac{\partial\phi}{\partial t} = \nabla\cdot(\mathbf{D}\cdot\nabla\phi) \tag{5.84}$$

Other forms: The concentration can be defined in three forms, the molar form ϕ_i with the unit $\frac{mol}{m^3}$, the volume form c_i with the unit $\frac{kg}{m^3}$, and the dimensionless mass fraction form $\Phi_i = \frac{c_i}{\sum_j c_i}$ with the unit $\frac{kg}{kg}$. Using the volume version of the concentration, Fick's first law is rewritten as

$$\frac{\dot{\mathbf{m}}_i}{A} = -D\nabla c_i \tag{5.85}$$

where $\dot{m}$ is the mass flux. Using the definition $\Phi = \frac{c_i}{\sum_i c_i}$

$$\nabla \ln \Phi_i = \frac{\nabla c_i}{c_i} \rightarrow \frac{\dot{m}}{A} = -c_i D \nabla (\ln \Phi_i) \tag{5.86}$$

Based on the relation $\dot{m}_i = c_i u_i A$, the velocity induced by mass diffusion is

$$\mathbf{u}_i = -D\nabla(\ln \Phi_i) \tag{5.87}$$

The general form of the diffusion velocity also contains the gradients of pressure and temperature,

$$\mathbf{u}_i = -D\nabla(\ln \Phi_i) - D_p \nabla(\ln p) - D_T \nabla(\ln T) \tag{5.88}$$

Two new terms may be ignored if the diffusivity coefficients due to pressure or temperature difference are smaller than the mass diffusivity coefficient or, respectively if $\nabla(\ln p) << \nabla(\ln \Phi_i)$ and $\nabla(\ln T) << \nabla(\ln \Phi_i)$.

5.4.2 MASS TRANSPORT EQUATIONS

Mass transfer for fluid flow: In this section, we are going to write the conservation of mass for a mixture of two or more fluids. Consider the two-dimensional element shown in Figure 5.2, and write down the mass conservation for this element,

$$
\begin{aligned}
\frac{\partial c_i}{\partial t} dx dy dz \;=\;& (c_i u_i dy dz) - \left(c_i u_i + \frac{\partial}{\partial x}(c_i u_i) dx \right) dy dz \\
+\;& (c_i v_i dx dz) - \left(c_i v_i + \frac{\partial}{\partial y}(c_i v_i) dy \right) dx dz \\
+\;& (c_i w_i dx dy) - \left(c_i w_i + \frac{\partial}{\partial z}(c_i w_i) dz \right) dx dy + \dot{m}_{si} dx dy dz
\end{aligned}
\tag{5.89}
$$

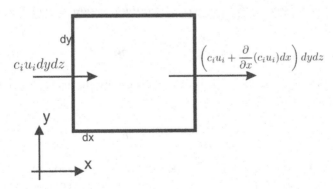

Figure 5.2 The two-dimensional element used to derive the conservation of mass in the mass transfer context.

where $\dot{m}_{si}$ is the volumetric rate of generation of constituent i ($\frac{kg}{m^3 s}$). By canceling similar terms, and dividing both sides by the volume of the element, we obtain

$$\frac{\partial c_i}{\partial t} + \frac{\partial}{\partial x}(c_i u_i) + \frac{\partial}{\partial y}(c_i v_i) + \frac{\partial}{\partial z}(c_i w_i) = \dot{m}_{si} \tag{5.90}$$

Add the equations written for all constituents. Then, the summation over i appears

$$\frac{\partial c}{\partial t} + \frac{\partial}{\partial x}\sum_i c_i u_i + \frac{\partial}{\partial y}\sum_i c_i v_i + \frac{\partial}{\partial z}\sum_i c_i w_i = \sum_i \dot{m}_{si} \tag{5.91}$$

Define $(u, v, w) = (\frac{1}{c}\sum_i c_i u_i, \frac{1}{c}\sum_i c_i v_i, \frac{1}{c}\sum_i c_i w_i)$ as the mass-averaged velocity components. It should be noted that the mass-averaged velocities may become zero even if the velocity of each constituent is not zero. Replacing $c = \sum_i c_i$ by ρ, we reach the familiar form of the mass conservation law,

$$\frac{\partial \rho}{\partial t} + \frac{\partial}{\partial x}(\rho u) + \frac{\partial}{\partial y}(\rho v) + \frac{\partial}{\partial z}(\rho w) = \dot{m}_s \tag{5.92}$$

Let's continue the process from Equation (5.90). Define the diffusion fluxes as $\dot{\mathbf{m}}_i = c_i(\mathbf{u}_i - \mathbf{u})$, which represent the flow rate of constituent i per unit area relative to the bulk motion of the mixture,

$$\frac{\partial c_i}{\partial t} + \frac{\partial}{\partial x}(c_i u) + \frac{\partial}{\partial y}(c_i v) + \frac{\partial}{\partial z}(c_i w) = -\frac{\partial \dot{m}_{xi}}{\partial x} - \frac{\partial \dot{m}_{yi}}{\partial y} - \frac{\partial \dot{m}_{zi}}{\partial z} + \dot{m}_{si}$$

$$\frac{\partial c_i}{\partial t} + \nabla \cdot (c_i \mathbf{u}) = -\nabla \cdot \dot{\mathbf{m}}_i + \dot{m}_{si} \tag{5.93}$$

This is the conservation form of the concentration transport equation. Using the conservation of mass $\frac{\partial c_i}{\partial t} + \nabla \cdot \mathbf{u} = 0$ and the rule for the derivative of product, the non-conservation form of the mass conservation law becomes

$$\frac{\partial c_i}{\partial t} + u\frac{\partial c_i}{\partial x} + v\frac{\partial c_i}{\partial y} + w\frac{\partial c_i}{\partial z} = -\frac{\partial \dot{m}_{xi}}{\partial x} - \frac{\partial \dot{m}_{yi}}{\partial y} - \frac{\partial \dot{m}_{zi}}{\partial z} + \dot{m}_{si}$$

$$\frac{Dc_i}{Dt} = -\nabla \cdot \dot{\mathbf{m}}_i + \dot{m}_{si} \tag{5.94}$$

Using Fick's first law and assuming constant diffusivity, the final advection-diffusion equation for mass transfer reads

$$\frac{Dc_i}{Dt} = D\nabla^2 c_i + \dot{m}_{si} \tag{5.95}$$

Phase separation: To model the process of phase separation, the Cahn–Hilliard equation can be used, which predicts the separation of two components of a binary fluid or alloy and the formation of domains pure in each component [120]. A similar equation in the context of phase separation in multi-component systems is the Allen–Cahn equation [121]

Mixing process: The mixing process happens based on microscopic diffusion across the interface between two adjacent media. We can strengthen the mixing process by mechanically stirring the media and including the advection terms in the transport equation. These two mechanisms are governed by an advection-diffusion equation for the concentration of the mixture in the dimensionless form (* refers to the normalized concentration)

$$\frac{Dc_m^*}{Dt} = \frac{1}{Pe_D}\nabla^2 c_m^* \tag{5.96}$$

where $Pe_D = \frac{UL}{D}$ is the diffusion Peclet number defined based on the diffusion coefficient.

Conservation of salinity: The salt content of oceans called the salinity changes point by point inside water due to local mixing, mass transfer from water free-surface, precipitation, or the phase change of the seawater. Since the total amount of salt in sea/ocean waters does not change, we may find a conservation law for the salinity of water (sometimes called the halinity). If D is the mass diffusivity of salt in water and s is the concentration of salt in water with the unit of gram of salt in kilogram of the saline water, the conservation of salinity reads

$$\frac{\partial(\rho s)}{\partial t} + \frac{\partial(\rho s u_j)}{\partial x_j} = -\frac{\partial \dot{m}_j}{\partial x_j}$$

$$\dot{m}_j = -D\frac{\partial(\rho s)}{\partial x_j} \tag{5.97}$$

Replacing Fick's law in the conservation of salinity,

$$\frac{\partial(\rho s)}{\partial t} + \frac{\partial(\rho s u_j)}{\partial x_j} = \frac{\partial}{\partial x_j}\left(D\frac{\partial(\rho s)}{\partial x_j}\right) \tag{5.98}$$

Similar to most other conservation laws, this is also an advection-diffusion equation involving the familiar material derivative operator. It should be noted that this equation is coupled with the continuity and the momentum equations via the density and velocity variations. This was previously called the s-compressibility.

The Boussinesq approximation neglects the variation of density in water and supposes the flow of water to be divergence-free. However, the velocity field of such flows is not divergence-free and the compressible form of the Navier–Stokes equation should be considered. Pay attention that in applications such as large-scale ocean flows, the Coriolis acceleration usually has to be considered.

However, the system of equations is not closed yet. Similar to the p-compressible flows, we need a relation between salinity, temperature, and density which is called the equation of state $\rho = f(T,s,p)$. One of the simplest equations of state for flows with slight density variations is

$$\rho = \rho_0[1 - \beta(T - T_0) + \alpha(s - s_0) + \gamma(p - p_0)] \tag{5.99}$$

where the subscript 0 refers to the reference state, $\beta \simeq 2 \times 10^{-4} k^{-1}$ is the thermal expansion coefficient, $\alpha \simeq 7.6 \times 10^{-4} ppt^{-1}$ is the saline contraction coefficient that encounters the effect of salinity change on density. Its value is one order of magnitude greater than the thermal expansion coefficient. $\gamma \simeq 4.1 \times 10^{-10} Pa^{-1}$ is the compressibility coefficient and equals the inverse of bulk modulus [122].

Homogenous reacting flows: If a chemical reaction occurs inside a flow or at the interface of a fluid with another fluid/solid, we call such cases reactive flows. Examples are combustion, burners, fuel cells, batteries, and reactors. Single-phase reacting flow can be regarded as a mixture of constituent species. So, reactive flows may be an instance of mass transfer problems. In this strategy, the constituent j is modeled as a continuum phase, which can chemically and hydrodynamically communicate with other species.

· **Chemical reaction:** Generally, R simultaneous chemical reactions can be written in the following form:

$$0 \rightleftharpoons \sum_{j=1}^{N} v_{jk} A_j; \quad k = 1, ..., R \quad (5.100)$$

where N is the number of species in reactions, A_j represents the constituent, and v_{jk} is the number of moles of species or the stoichiometric coefficient. In this form, all constituents are written on one side of the reaction, and therefore, v_{jk} is negative for reactants. Consider that indices j and k in the relations are simple counters and have nothing to do with tensor notation rules. They just determine the chemical reaction corresponding to the constituent j in reaction k.

For example, assume that the reversible reaction $3Fe(s) + 4H_2O(g) \rightleftharpoons Fe_3O_4(s) + 4H_2(g)$ is one reaction from a set of five reactions ($k = 5$). This reaction can be written as $0 \rightleftharpoons Fe_3O_4(s) + 4H_2(g) - 3Fe(s) - 4H_2O(g)$. Then, the coefficients v_{j5}, j from 1 to $N = 4$ are 1, 4, −3, −4. If the reactants have different phases like solid, gas, or liquid, the reaction is heterogenous. Such heterogenous reactive flows should be simulated using multiphase flow models. Conversely, the homogenous reactions only contain a mixture of species in one phase.

· **Continuity and mass transport:** Chemical reactions change the mass density of the constituent j, based on the following conservation of mass relation

$$\frac{\partial \rho_j}{\partial t} + \nabla \cdot (\rho_j \mathbf{u}_j) = W_j \quad (5.101)$$

where W_j is the volumetric mass production rate of the constituent j via chemical reaction based on the Arrhenius reaction rate concept,

$$W_j = m_j \sum_{k=1}^{N} v_{kj} w_k \quad (5.102)$$

where m_j is the molecular weight of the constituent j, and w_k is the rate of reaction k among total R reactions. The reaction rate is the ratio of variation of concentration of species and their stoichiometric coefficients. It should be mentioned that the stoichiometric coefficients for reactants should be negative numbers.

Performing the summation over $j = 1, ..., N$ for all N species and considering $\rho = \sum_{j=1}^{N} \rho_j$, $\rho\mathbf{u} = \sum_{j=1}^{N} \rho_j\mathbf{u}_j$, and $\sum_{j=1}^{N} W_j = 0$, the familiar version of the conservation of mass appears,

$$\frac{\partial \rho}{\partial t} + \nabla \cdot (\rho\mathbf{u}) = 0. \tag{5.103}$$

$\sum_{j=1}^{N} W_j = 0$ is expected since the mass exchange between different species is an internal process. The same total amount of the produced mass should be absorbed by other species. Defining the mass fraction $Y_j = \frac{\rho_j}{\rho}$ and the diffusion flux vector $\mathbf{J}_j = \rho_j(\mathbf{u}_j - \mathbf{u})$, the final form of the local mass transfer equation is

$$\rho \frac{DY_j}{Dt} + \nabla \cdot \mathbf{J}_j = W_j. \tag{5.104}$$

$\mathbf{J}_j = -D_j \nabla(\rho Y_j)$ is given by Fick's law of diffusion, where D_j is the mass diffusivity of the constituent j in an R-reaction system.

· **The energy equation:** The modified version of the Fourier's law including the enthalpy flux is

$$\mathbf{q} = -k\nabla T + \sum_{i=1}^{N} (h_i \mathbf{J}_i) \tag{5.105}$$

Hence, the energy equation for reactive flows is

$$\rho c_p \frac{DT}{Dt} = \frac{Dp}{Dt} + \nabla \cdot (k\nabla T) - \sum_{i=1}^{N} (c_{pi}\mathbf{J}_i \cdot \nabla T) - \sum_{i=1}^{N} (h_i W_i) + Q_R + \Phi \tag{5.106}$$

where c_{pi}, Q_R, h_i are the specific heat of constituent i, the radiative heat transfer term, and the enthalpy of constituent i [123]. The material derivative of pressure can be approximated by its local derivative when the propagation of acoustic pressure waves is not important. The local time derivative may also be neglected in applications with constant pressure such as open flames [124]. The viscous dissipation term is negligible in low-speed flows. To close the equations, we need to compute the conductivity and viscosity of the mixture, which are equal to the sum of arithmetic and harmonic means of the conductivity and viscosity of each species [125]. For high-speed and high-Mach compressible flows, an appropriate equation of state should be added to the set of governing equations. A well-known simple equation of state for a perfect mixture of gases is $p = \frac{\rho RT}{W}$, where W is the mean molecular weight of the mixture [126].

5.5 CONSERVATION/NON-CONSERVATION FORMS OF LAWS

The conservation form of mass, momentum, energy and concentration conservation laws are

$$\frac{\partial \mathbf{U}_i}{\partial t} + \frac{\partial \mathbf{F}_i}{\partial x_i} + \frac{\partial \mathbf{G}_i}{\partial x_i} = \mathbf{B}_i \qquad (Q5.5) \quad (5.107)$$

The first term is the local time-derivative, the second term represents the convection fluxes, the third term shows the diffusion fluxes, and the last term denotes the source term,

$$\begin{aligned}
\mathbf{U} &= \begin{bmatrix} \rho & \rho u_j & \rho e' & c \end{bmatrix}^T \\
\mathbf{F}_i &= \begin{bmatrix} \rho u_i & \rho u_i u_j + p\delta_{ij} & \rho e' u_i + p u_i & c u_i \end{bmatrix}^T \\
\mathbf{G}_i &= \begin{bmatrix} 0 & -\tau_{ij} & -\tau_{ij} u_j + q_i & \dot{m}_i \end{bmatrix}^T \\
\mathbf{B} &= \begin{bmatrix} 0 & F_j & S + F_j u_j & \dot{m}_s \end{bmatrix}^T
\end{aligned} \qquad (5.108)$$

where $e' = e + \frac{u^2}{2}$. For example, the conservation form of the continuity equation in cartesian coordinates is

$$\frac{\partial \rho}{\partial t} + \frac{\partial}{\partial x}(\rho u) + \frac{\partial}{\partial y}(\rho v) + \frac{\partial}{\partial z}(\rho w) = 0 \qquad (5.109)$$

Expanding the derivative of the product,

$$\frac{\partial \rho}{\partial t} + \rho \frac{\partial u}{\partial x} + u \frac{\partial \rho}{\partial x} + \rho \frac{\partial v}{\partial y} + v \frac{\partial \rho}{\partial y} + \rho \frac{\partial w}{\partial z} + w \frac{\partial \rho}{\partial z} = 0 \qquad (5.110)$$

The definition of the material derivative operator yields

$$\frac{D\rho}{Dt} + \rho \nabla \cdot \mathbf{u} = 0 \qquad (5.111)$$

which is the non-conservation form of the continuity equation. Similarly, the conservation form of the x-momentum equation in cartesian coordinates is

$$\frac{\partial}{\partial t}(\rho u) + \frac{\partial}{\partial x}(\rho u^2) + \frac{\partial}{\partial y}(\rho vu) + \frac{\partial}{\partial z}(\rho wu) = -\frac{\partial p}{\partial x} + \frac{\partial \tau_{xx}}{\partial x} + \frac{\partial \tau_{yx}}{\partial y} + \frac{\partial \tau_{zx}}{\partial z} + F_x$$
$$(5.112)$$

Again, by expanding the derivative of the product and subtracting the continuity equation, it is easy to show that the previously obtained form of the momentum equation is recovered. The conservation $e - V$ form of the energy equation reads

$$\frac{\partial}{\partial t}(\rho e') + \frac{\partial}{\partial x}(\rho e' u) + \frac{\partial}{\partial y}(\rho e' v) + \frac{\partial}{\partial z}(\rho e' w) = -\frac{\partial}{\partial x}(pu) - \frac{\partial}{\partial y}(pv) - \frac{\partial}{\partial z}(pw)$$
$$+ \frac{\partial}{\partial x}(\tau_{xx} u + \tau_{xy} v + \tau_{xz} w) + \frac{\partial}{\partial y}(\tau_{yx} u + \tau_{yy} v + \tau_{yz} w) + \frac{\partial}{\partial z}(\tau_{zx} u + \tau_{zy} v + \tau_{zz} w)$$
$$- \left(\frac{\partial q_x}{\partial x} + \frac{\partial q_y}{\partial y} + \frac{\partial q_z}{\partial z} \right) + (F_x u + F_y v + F_z w) + S \qquad (5.113)$$

It is a good exercise to show that after expanding the derivatives and subtracting a continuity and a momentum equation, the previously mentioned form of the energy equation can be recovered.

The conservation form of the second law of thermodynamics is obtained by adding the product of s and the conservation of mass to the non-conservation form of the second law of thermodynamics,

$$\rho \frac{Ds}{Dt} \geq - \nabla \cdot \left(\frac{\mathbf{q}}{T}\right) + \frac{S}{T} \rightarrow$$

$$\frac{\partial(\rho s)}{\partial t} + \nabla \cdot (\rho \mathbf{u} s) + \nabla \cdot \left(\frac{\mathbf{q}}{T}\right) - \frac{S}{T} \geq 0 \qquad (5.114)$$

Remarks

- It should be noted that the conservation and non-conservation forms of equations are mathematically equivalent. Also, both of them have to satisfy conservation laws. The non-conservation title refers to probable numerical problems during simulations, which may lead to the violation of the conservation laws. If we use the non-conservation form of equations for compressible flows or in the finite-volume numerical method, then the governing equations may not truly be satisfied. For instance, crossing a shock wave in supersonic flows, the velocity, the density, the pressure, and the temperature have discontinuities. But in the conservation form of laws, combinations of such quantities create physical fluxes without experiencing a discontinuous behavior.
- In compressible flows and numerical methods like the finite-volume, it is essential to use the conservation form of equations. In other cases, it is recommended to use the conservation form of the laws.
- If $\mathbf{G}_i = \mathbf{0}$, the Euler equation is obtained.
- The conservation form contains the divergence of a flux, but the non-conservation form includes the material derivative term. So, the other name for the conservation form is the "divergence form".
- Integrating both sides of Equation (5.107) over volume, we can obtain the control volume form of the governing equations,

$$\int_{C.V.} \left(\frac{\partial \mathbf{U}_i}{\partial t} + \frac{\partial \mathbf{F}_i}{\partial x_i} + \frac{\partial \mathbf{G}_i}{\partial x_i} - \mathbf{B}_i\right) d\forall = 0 \qquad (5.115)$$

and with the use of the divergence theorem,

$$\int_{C.V.} \left(\frac{\partial \mathbf{U}_i}{\partial t} - \mathbf{B}_i\right) d\forall + \int_{C.S.} (\mathbf{F}_i + \mathbf{G}_i) n_i dA = 0 \qquad (5.116)$$

The integral form of the second law of thermodynamics for control volume can be obtained by integrating Equation (5.114) over volume. After interchanging the derivative and the integral and using the divergence theorem

$$\frac{\partial}{\partial t} \int_{C.V.} \rho s d\forall + \int_{C.S.} \rho s (\mathbf{u} \cdot \mathbf{n}) dA + \int_{C.S.} \frac{\mathbf{q} \cdot \mathbf{n}}{T} dA - \int_{C.V.} \frac{S}{T} d\forall \geq 0 \quad (5.117)$$

5.6 DIMENSIONLESS FORM OF LAWS

Using the dimensionless groups defined based on characteristic parameters shown by the subscript c,

$$t^* = \frac{t}{t_c}, \quad x_i^* = \frac{x_i}{L_c}, \quad g_i^* = \frac{g_i}{g_c}, \quad \Phi^* = \frac{L_c^2}{\mu_c U_c^2}\Phi, \quad k^* = \frac{k}{k_c}, \quad s^* = \frac{sL_c \rho_c U_c}{k_c}$$

$$\rho^* = \frac{\rho}{\rho_c}, \quad u_i^* = \frac{u_i}{U_c}, \quad \mu^* = \frac{\mu}{\mu_c}, \quad T^* = \frac{\Delta T}{\Delta T_c}, \quad c_p^* = \frac{c_p}{c_{pc}}, \quad p^* = \frac{\Delta p}{p_c}$$

$$(5.118)$$

the following normalized forms of the conservation laws will be obtained. The conservation of mass,

$$\left(St\frac{\partial \rho^*}{\partial t^*} + u^*\frac{\partial \rho^*}{\partial x^*} \right) + \rho^*\frac{\partial u_i^*}{\partial x_i^*} = 0 \qquad (5.119)$$

the momentum conservation law,

$$\rho^*\left[St\frac{\partial u_j^*}{\partial t^*} + u_i^*\frac{\partial u_j^*}{\partial x_i^*} \right] = -Eu\frac{\partial p^*}{\partial x_j^*} + \frac{1}{Re}\frac{\partial}{\partial x_i^*}\left[\mu^*\left(\frac{\partial u_j^*}{\partial x_i^*} + \frac{\partial u_i^*}{\partial x_j^*} \right) - \frac{2}{3}\delta_{ij}\mu^*\frac{\partial u_k^*}{\partial x_k^*} \right]$$

$$+ \frac{1}{Fr}\rho^* g_j^* \qquad (5.120)$$

the energy equation,

$$\rho^* c_p^*\left[St\frac{\partial T^*}{\partial t^*} + u_i^*\frac{\partial T^*}{\partial x_i^*} \right] = Ec\frac{Dp^*}{Dt^*} + \frac{1}{RePr}\nabla\cdot(k^*\nabla T^*) + \frac{Ec}{Re}\Phi^* \qquad (5.121)$$

the entropy equation,

$$\rho^*\left[St\frac{\partial s^*}{\partial t^*} + u_i^*\frac{\partial s^*}{\partial x_i^*} \right] = -\nabla\left(\frac{\mathbf{q}^*}{T^*} \right) - \frac{1}{T^{*2}}\mathbf{q}^*\cdot\nabla T^* + EcPr\frac{\Phi^*}{T^*} \qquad (5.122)$$

and the equation of state for ideal gases with $\rho_c U_c^2$ as the pressure scale,

$$p = \rho RT \rightarrow (M^2\gamma)p^* = \rho^* T^* \qquad (5.123)$$

The relation for the speed of sound for calorically perfect gas is $c^2 = \gamma RT$. $PrEc$ is another dimensionless group called the Brinkman number. In all relations, the del operator has been nondimensionalized based on the characteristics length (L_c) and the subscript * for ∇ has been dropped for the sake of simplicity. The following dimensionless groups appear in the above-mentioned conservation laws,

$$Fr = \frac{U_c^2}{g_c L_c}, \quad St = \frac{L_c}{t_c U_c}, \quad Pr = \frac{\mu_c c_p}{k_c},$$

$$Eu = \frac{\Delta p_c}{\rho_c U_c^2}, \quad Re = \frac{\rho_c U_c L_c}{\mu_c}, \quad Ec = \frac{U_c^2}{c_{pc}T_c} = (\gamma - 1)M^2 \qquad (5.124)$$

The product of $RePr$ in the energy equation is the Peclet number. If the pressure difference (Δp) is defined as the difference between the pressure and the vapor pressure, the Eu number (the coefficient of pressure) is replaced by the cavitation number, $Ca = \frac{p - p_v}{0.5\rho_c U_c^2}$ (Q5.4).

5.6.1 REMARKS

- **Normalization of the body force in natural convection:** The dimensionless numbers commonly used in natural convection are the Grashohf number, the Richardson number, and the Rayleigh number. The Richardson number is used to determine the mixed/forced/natural convection regime, while the Rayleigh number determines the conduction/laminar/turbulent natural convection regime. The Grashohf number is defined as the ratio of the driving force and the resisting force similar to the structure of definition of the Reynolds number in forced convection.

$$Gr = \frac{g\beta\rho_c^2 L_c^3 (T_w - T_c)}{\mu_c^2} = \frac{Buoyancy\ effects}{Viscous\ effects}$$

$$Ri = \frac{Gr}{Re^2} = \frac{Buoyancy\ effect}{Inertia\ effect}$$

$$Ra = GrPr \tag{5.125}$$

The dimensionless body force due to natural convection in the Navier–Stokes equations can be written as $\frac{Gr}{Re^2}T^* = RiT^*$.

- **Normalization of the boundary layer equations:** If we choose different length-scales in x and y directions based on the integral solution of the laminar boundary layer equations, $x^* = \frac{x}{L_c}$, $y^* = \frac{y}{L_c}\sqrt{Re}$ and similarly for the velocity components $u^* = \frac{u}{U_c}$, $v^* = \frac{v}{U_c}\sqrt{Re}$, the Reynolds number vanishes from the momentum equations,

$$0 = \frac{\partial u^*}{\partial x^*} + \frac{\partial v^*}{\partial y^*}$$

$$\frac{\partial u^*}{\partial t^*} + u^*\frac{\partial u^*}{\partial x^*} + v^*\frac{\partial u^*}{\partial y^*} = -\frac{\partial p^*}{\partial x^*} + \frac{\partial^2 u^*}{\partial y^{*2}}$$

$$-\frac{u^{*2}}{\mathfrak{R}^*} = -\frac{\partial p^*}{\partial y^*} \tag{5.126}$$

where x and y are parallel-to and normal-to-wall coordinates (Q5.6).

- **Normalization of Stokes flow:** When the inertia is negligible, we need to use another characteristic pressure called the viscous pressure ($\frac{\mu U_c}{L_c}$) to nondimensionalize pressure,

$$p^* = \frac{\Delta p}{\frac{\mu U_c}{L_c}} \tag{5.127}$$

The Poiseuille number $Po = C_f Re$ can be defined in non-inertial (creeping) flows like fully-developed duct flows.

- **The non-Fourier moving media:** A special type of non-dimensionalization has been introduced (Khayat et al, 2015) [91] by selecting $\frac{\alpha}{L}$, $\frac{U}{L}$, $\frac{\mu U}{L}$, $-\alpha \nabla T$ as velocity-, time-, stress-, and heat flux-scale, respectively,

$$Pr^{-1}\frac{D\mathbf{u}^*}{Dt} = -\nabla p^* + \nabla \cdot \boldsymbol{\tau}^*$$

$$C\left(\frac{\partial \mathbf{q}^*}{\partial t^*} + \mathbf{u}^* \cdot \nabla \mathbf{q}^* - \mathbf{q}^* \cdot \nabla \mathbf{u}^*\right) + \mathbf{q}^* = -\nabla T^* \qquad (5.128)$$

where C is the Cattaneo number. This way, the Prandtl number appears in the momentum equation.

- **Normalization of the inertial term:** The dimensionless form of the Coriolis acceleration contains the Rossby number based on the angular velocity of the rotating frame (Ω_c),

$$\mathbf{a}^*_{Coriolis} = \frac{1}{Ro}(2\boldsymbol{\Omega}^* \times \mathbf{u}^*)$$

$$Ro = \frac{U_c}{\Omega_c L_c} \qquad (5.129)$$

The Taylor number is the ratio of the centripetal acceleration to the viscous force, $Ta = \frac{4\omega^2 R_c^4}{\nu^2}$.

- **Normalization of boundary conditions:** The dimensionless form of Maxwell's slip-condition, and the temperature-jump condition includes the Knudsen number, $Kn = \frac{\lambda}{L_c}$,

$$u_s^* = Kn\left(\frac{\partial u^*}{\partial n^*}\right)_w$$

$$T_w^* = 1 + \frac{2Kn\gamma}{Pr(\gamma+1)}\left(\frac{\partial T^*}{\partial n^*}\right)_w$$

$$\gamma = \frac{c_p}{c_v}$$

The normalized form of the convection boundary condition reads

$$Nu = \frac{hL_c}{k_c} = \left(k^*\frac{\partial T^*}{\partial n^*}\right)_w \qquad (5.130)$$

Similarly, the Sherwood number is defined as $Sh = \frac{h_D L_c}{D_c}$, based on the convection coefficient of mass transfer (h_D) and the mass diffusivity coefficient (D_c).

The dimensionless surface tension term in normal-to-wall stress balance (Equation 4.104) is connected to the inverse of the Weber number. Using the Stokes constitutive relation for incompressible flows,

$$-p^* + \frac{2}{Re}\mathbf{n} \cdot \boldsymbol{\varepsilon}^* \cdot \mathbf{n} = \frac{1}{We}(\nabla \cdot \mathbf{n}) \qquad (5.131)$$

where $We = \frac{\rho_c U_c^2 L_c}{\sigma}$, and $\boldsymbol{\varepsilon}^*$ is the dimensionless strain-rate tensor.

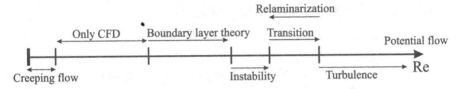

Figure 5.3 Different flow regimes on the Reynolds axis.

- **Normalization for nanoscale conduction problems:** Here, we use the relation $k = \frac{1}{3}Cv\lambda$ in which C and v are the specific heat per unit volume and the average velocity of sound in solids, respectively. Defining $t^* = \frac{t}{\tau_T}$, $B = \frac{\tau_T}{\tau_q}$, the dimensionless form of the energy equation is obtained containing the Knudsen number [96]

$$\frac{\partial T^*}{\partial t^*} + \frac{\partial^2 T^*}{\partial t^{*2}} = \frac{Kn^2}{3}\nabla^2 T^* + B\frac{Kn^2}{3}\frac{\partial}{\partial t^*}\nabla^2 T^* \qquad (5.132)$$

In this manner, the Knudsen number appears not only in boundary conditions, but also inside the equation.
- Dimensionless numbers are commonly used to categorize different flow conditions. For instance, the Mach number classifies compressible flow regimes, the Reynolds number classifies viscous flow regimes, the Froude number specifies free-surface flow regimes, the Knudsen number determines the continuum mechanics regimes, the Stokes number and the density ratio imply the two-phase flow regimes and so on.

Based on the definition of the Reynolds number, the Reynolds axis has been sketched in Figure 5.3. Increasing the Reynolds number from zero to infinity, we have the following regimes.

1. The creeping flow without acceleration. In creeping internal flows, the assumption of zero acceleration can be extended to high Reynolds numbers. The Stokes paradox for external two-dimensional flows becomes important in this region.
2. The moderate Reynolds number region that can be treated just by computational fluid dynamics and the solution of the full Navier–Stokes equations.
3. The laminar boundary layer flow with the Reynolds numbers approximately above 1000.
4. Unstable laminar flows.
5. Transition to turbulence or the reversed process: relaminarization.
6. Fully turbulent flows (turbulent boundary layer approximation).
7. Potential flow with the well-known d'Alembert paradox.

5.6.2   LIST OF DIMENSIONLESS NUMBERS (Q5.7)

- **Preprocessing numbers**

  1. Geometric dimensionless numbers.
  2. Specific gravity: $SG = \frac{\rho_c}{\rho_r}$ where ρ_r is the reference density referring to water for solids/liquids and to air for gases.
  3. Relative roughness: $e = \frac{\varepsilon}{L_c}$ appearing in turbulent flows.
  4. Specific heat ratio or isentropic index: $\gamma = \frac{c_p}{c_v}$.
  5. Minor loss coefficient in duct flows which equals 1 and 2 for uniform flow and laminar Poiseuille profile entering an infinite reservoir, respectively.

- **Post-processing numbers**

  1. Lift coefficient: $C_L = \frac{F_L}{0.5\rho_c U_c^2 L_c^2}$.
  2. Drag coefficient: $C_D = \frac{F_D}{0.5\rho_c U_c^2 L_c^2}$.
  3. Darcy (f) and skin/fanning friction coefficients ($C_f = f/4$): $C_f = \frac{\tau_w}{0.5\rho_c U_c^2}$.
  4. Torque coefficient: $C_M = \frac{T}{0.5\rho_c U_c^2 L_c^3}$.
  5. Power coefficient: $C_P = \frac{P}{0.5\rho_c U_c^3 L_c^2}$.
  6. Tip-speed ratio (TSR): It appears in the performance evaluation of wind turbines.
  7. Efficiency η.
  8. Deci Bell dB: as a logarithmic scale appears in acoustic applications.
  9. Head coefficient: $\psi = \frac{gH}{U^2}$ where U is the circumferential velocity of turbomachinery.
  10. Flow coefficient: $\phi = \frac{C_m}{U}$ where C_m is the radial velocity of the moving fluid between blades of the impeller.

- **Fluid mechanics**

  1. Reynolds number: $Re = \frac{\rho U_c L_c}{\mu}$ is proportional to the ratio of inertial forces to viscous forces appearing in all viscous flows. Bejan (2013) [97] stated that the Reynolds number can be interpreted as a geometric parameter in boundary layers. So the fact that we use the Reynolds number in flows without inertia can be justified. In other flows without acceleration, the Re number is replaced by the Po number.
  2. Euler number, pressure coefficient, cavitation number: Eu, C_p, $Ca = \frac{\Delta p}{p_c}$ are the ratio of pressure forces to inertial forces (in flows with inertia) or the viscous pressure (in Stokes flows).
  3. Cauchy number or Mach number: $M = \sqrt{Ca} = \frac{U_c}{c_c}$ is proportional to the ratio of kinetic energy to internal energy.
  4. Froude number: $Fr = \frac{U_c}{\sqrt{L_c g_c}}$ is the ratio of inertial forces to the gravity force. It appears in free-surface flows and boiling.

5. Richardson number: $Ri = \frac{1}{Fr^2}$ the ratio of potential energy to kinetic energy. It appears in atmospheric and oceanic flows, natural convection, and flow instabilities.

6. Galileo number and Archimedes number: $Ga, Ar = Re^2 Ri = \frac{gL_c^3}{\nu^2}$, the ratio of the gravity force to viscous forces.

7. Contact and sliding angles which appear in surface tension phenomenon.

8. Weber number: $We = \frac{\rho U_c^2 L_c}{\sigma}$ is the ratio of inertial forces to the surface tension force. It appears in surface tension-involved flows such as boiling, cavitation, bubble dynamics, drop collision, jets, and free-surface flows.

9. Ohnesorge number: $Oh = \frac{\sqrt{We}}{Re}$.

10. Marangoni number: $Ma = \frac{L_c \Delta \sigma}{L \sigma}$ where $\Delta\sigma/L$ represents the characteristic stress in Marangoni convection.

11. Capillary number: $Ca = \frac{We}{Re} = \frac{\mu U_c}{\sigma}$ is the ratio of viscous forces to the surface tension force.

12. Laplace number: $La = \frac{Re^2}{We} = \frac{\sigma \rho L_c}{\mu^2}$ is the ratio of the surface tension force to the rate of momentum transfer due to viscous forces.

13. Eötvös number and Bond number: $Bo, Eo = \frac{We}{Fr^2} = \frac{\rho_c g_c L_c^2}{\sigma}$, the ratio of the buoyancy force to the surface tension force.

14. Morton number: $Mo = \frac{We^3}{FrRe^4} = \frac{g_c \mu_c^4}{\rho_c \sigma^3}$.

15. Thoma's cavitation factor: $\sigma = \frac{NPSH}{H}$ is the ratio of the net positive suction head to the head of the pump/turbine.

16. Strouhal number: $St = \frac{L_c}{U_c t_c}$ is the ratio of the local inertial term to convective inertia. It appears in frequency-involved problems such as vortex shedding, aeroacoustics, and oscillations [98].

17. Keulegan–Carpenter number: $K_c = \frac{1}{St}$.

18. Womersley number: $\alpha = \sqrt{2\pi Re St}$ is the ratio of time-dependent inertial forces to viscous forces. It appears in periodic biological flows such as blood flow in the aorta.

19. Roshko number: $Ro = ReSt$.

20. Hartmann number: $Ha = B_c L_c \sqrt{\frac{\sigma_e}{\mu_c}}$ is the ratio of the Lorentz force to the viscous force. B_c and σ_e are the magnetic field and the electrical conductance of the fluid.

21. Stuart number: $N = \frac{B_c^2 L_c \sigma_e}{\rho_c U_c}$ is the ratio of electromagnetic forces to inertial forces.

22. Shape factor: $H = \frac{\delta^*}{\theta}$ is the ratio of the displacement thickness to the momentum thickness in boundary layer theory.

23. Bejan number in fluid mechanics: $Be_F = \frac{\Delta p L_c^2}{\mu_c \nu}$.

24. Deborah number: $De = \frac{t_c}{t_p}$ in which t_c is the relaxation time and determines the degree of solid-like behavior of viscoelastic materials.

25. Weissenberg number: $Wi = \varepsilon_l t_c$ is the ratio of the elastic force to the viscous force. Wi indicates the degree of anisotropy or orientation generated by deformation.

26. Elasticity number: $E = \frac{t_c \nu}{L_c^2}$ is the ratio of elastic forces to inertial forces. It appears in the modeling of non-Newtonian viscoelastic fluids.

27. Bingham number: $Bm = \frac{\tau_c L_c}{\mu_c U_c}$ is the ratio of the yield stress to the viscous stress. It appears in the rheology of Bingham fluids.

28. Rossby number: $Ro = \frac{U_c}{\Omega_c L_c}$ is the ratio of inertial forces to the Coriolis force. It appears in arctic engineering, turbomachinery, oceanic flows, and weather forecasting.

29. Ekman number: $Ek = \frac{Ro}{Re}$ is the ratio of viscous forces to the Coriolis force. It appears in large-scale geophysical flows.

30. Taylor number: $Ta = \frac{4\omega^2 R_c^4}{\nu^2}$ is the ratio of the centripetal acceleration to viscous forces.

31. Dean number: $De = Re\sqrt{\frac{L_c}{2\Re}}$ is the product of the Reynolds number and the ratio of the centripetal acceleration to the inertial force. It appears in flows inside curved pipes.

32. Sommerfeld number: $So = \frac{r^2}{c^2}\frac{\mu_c N}{P}$ in which r, c, N, P are the radius of the shaft, the radial clearance, the angular speed, and the load intensity. It appears in lubrication theory and analysis of bearings.

33. Knudsen number: $Kn = \frac{\lambda}{L_c}$ is the ratio of the mean-free-path to the characteristic length. It is important in non-continuum flows.

34. Darcy number: $Da = \frac{K}{L_c^2}$ is the ratio of permeability of the porous medium to the area scale.

35. Porosity: $\phi = \frac{V_V}{V_T}$ is the ratio of the volume of pores in a porous material to the total volume.

36. Courant number $C = \frac{u\Delta t}{\Delta x}$ appears in computational fluid dynamics as the CFL stability condition.

37. Karlovitz number: $Ka = \frac{t_F}{t_\eta}$ is the ratio of the chemical time-scale to the Kolmogorov time-scale. It appears in turbulent combustion.

38. Damköhler number: $Da = \frac{1}{Ka}$.

• **Heat and mass transfer**

1. Rayleigh number: $Ra = \frac{g_c \beta}{\nu \alpha}\Delta T L_c^3$ is the criterion for transition to turbulence in natural convection.

2. Prandtl number: $Pr = \frac{\nu}{\alpha}$ is the ratio of momentum diffusion to thermal diffusion.

3. Peclet number: $Pe = Pr Re$.

4. Graetz number: $Gz = Re_{D_H} Pr = \frac{\dot{m} c_p D_H}{k_c L_c^2}$ is the ratio of thermal capacity to conduction heat transfer. It appears in developing flows in conduits.

5. Grashof number: $Gr = Ra Pr$.

6. Bejan number in heat transfer: $Be_H = \frac{\Delta p L_c^2}{\mu_c \alpha}$.

7. Nusselt number: $Nu = \frac{hL_c}{K_f}$ is the ratio of convective to conductive heat transport.

8. Stanton number: $St = \frac{Nu}{RePr} = \frac{h}{\rho_c U_c c_p}$ appears in the momentum–thermal Reynolds analogy.

9. Sherwood number: $Sh = \frac{h_D L_c}{D}$ appears in mass transfer and is analogous to the Nusselt number in heat transfer.

10. Schmidt number: $Sc = \frac{v}{D}$ is the ratio of momentum and mass diffusivities.

11. Lewis number: $Le = \frac{Sc}{Pr} = \frac{\alpha}{D}$ is the ratio of thermal and mass diffusivities.

12. Stanton number for mass transfer: $St_D = \frac{Sh}{ReSc}$ appears in the mass–momentum Reynolds analogy.

13. Colburn J-factor: $J_H = StPr^{\frac{2}{3}}$ appears in the Chilton–Colburn analogy.

14. Biot number: $Bi = \frac{hL_c}{k_s}$ is the ratio of the strength of surface to volumetric heat transfer. It appears in transient conductive heat transfer.

15. Fourier number: $Fo = \frac{\alpha t_c}{L_c}$ is the ratio of conduction heat transfer to the stored heat. It appears in transient heat transfer as the normalized time.

16. Eckert number: $Ec = \frac{U_c^2}{c_p(T_c - T_w)}$ is the ratio of kinetic energy to enthalpy difference.

17. Brinkman number: $Br = \frac{\mu_c U_c^2}{k_c(T_x - T_c)} = PrEc$ is the ratio of viscous dissipation to conduction effects. It appears in biological and polymeric flows.

18. Cattaneo number: $C = \frac{\tau \alpha}{L^2}$ appears in non-Fourier heat transfer.

19. Jacob number: $Ja = \frac{c_p(T_c - T_w)}{h_{fg}}$ is the ratio of the sensible heat to the latent heat. It appears in phase-change boiling flows.

20. Kutateladze number: $K = \frac{U_c \sqrt{\rho_{l_c}}}{(\sigma g_c (\rho_l - \rho_g))^{0.25}}$ is the ratio of inertial forces to the surface tension and gravitational force.

21. Stephan number: $Ste = \frac{c_p(T_c - T_w)}{L_c}$ is the ratio of the sensible heat to the latent heat of melting. Appears in phase-change liquid–solid flows.

22. Bejan number in thermodynamics: $Be = \frac{\dot{S}_{gen,T}}{\dot{S}_{gen,T} + \dot{S}_{gen,P}}$ is the ratio of entropy generation due to temperature difference to total entropy generation.

23. Sparrow number: $\frac{hL^2}{kr}$ where h is an interstitial heat transfer coefficient that controls the local thermal equilibrium, L is the porous layer thickness, and r is the hydraulic radius [84].

Some other nondimensional numbers are the Zel'dorvich number, the Urshell number, the Hodgson number, the Mason number, the Blake number, the Markstein number, the Arrhenius number, the Atwood number, the Brownett number, and the Chandraselehar number.

5.7 A REVIEW OF PHYSICAL ASPECTS

During the history of continuum mechanics, some well-known paradoxes and analogies have been presented for heat transfer and fluid mechanics. As a researcher in related fields, we need to be familiar with such items as a part of our general knowledge. Let us start from paradoxes and go on with analogies.

Counterintuitive phenomena: Paradoxes are unexpected phenomena that are not compatible with the predictions of equations.

1. **The Stokes' paradox at small Reynolds numbers:** The two-dimensional Stokes flow cannot satisfy the boundary conditions of flow over bodies due to the appearance of a logarithmic singularity at the far field. In other words, a two-dimensional creeping flow solution is impossible.

2. **The D'Alembert paradox at high Reynolds limit:** In an inviscid incompressible flow over a bluff body of any shape, the net drag force is zero.

3. **The boundary layer paradox:** The vertical velocity obtained from the solution of the boundary layer equations in Blasius flow is nonzero, which is unexpected and may lead to the appearance of rotational flow in out-of-boundary layer region [79].

4. **The energy equation paradox:** Coefficient of the time derivative term in the energy equation may be c_p or c_v. For liquids and solids they are equal constants, but for gases, it matters which one should be used in the equation. Panton [79] discussed this point and indicated that the correct choice is c_p.

5. **The heat transfer paradox:** The speed of propagation of thermal waves is finite. However, Fourier's law yields infinite sound speed.

6. **The Pascal's paradox:** Suppose three containers with different shapes but identical base surfaces, are filled with a liquid to the same depth. Although the pressure distribution over the base surface and the circumferential areas are equal, the scale shows different weights for each case.

7. **The Knudsen's paradox:** Based on the predictions of the semi-continuum equations, it is expected that the mass flux of a gas streaming in a channel decreases with increasing the Knudsen number. However, a minimum mass flow rate is seen around the Knudsen number equal to 0.8 above which the flow rate starts to increase.

8. **The tea leaf paradox:** When you stir a cup of tea, the leaves move inward and gather together near the center of the glass. While it is expected that the centrifugal force will push them outward. This is due to the creation of secondary flows near the bottom of the cup, where the fluid velocity is nearly zero.

9. **The upstream contamination:** When a fluid is poured out of a higher container into a lower one, particles can travel upward and move toward the upstream.

10. **The Gibbs paradox:** Gibbs observed that the entropy of an ideal gas is not an extensive variable. This point leads to the violation of the second law of thermodynamics. The generalized version of the Gibbs paradox is the mixing paradox.

11. **The Loschmidt's paradox or the reversibility paradox** indicates that there is a paradox between the time-invariance of the governing equations and the second law of thermodynamics.
12. **The Maxwell's demon:** There are two containers filled with two gases and a wall between them. The key to the door is in the hands of a demon. The paradox says that there may be a clever condition under which the second law of thermodynamics is violated.
13. **The Mpemba effect** says that hot water freezes faster than cold water. Generally speaking, it states that bigger events have higher accelerations.

Analogies: The following items are the celebrated analogies between two diverse fields of science, which have similar mathematical relations. Similarities between the governing equations motivate us to predict the behavior of one system by imitating the results of the other.

1. **Acoustic analogy** as mentioned in Section 4.1.2.4.
2. **Stokes–Rayleigh analogy** between buckling of an elastic membrane (studied by Rayleigh) and a thin viscous film (studied by Stokes).
3. **Reynolds analogy** refers to similarity of the energy equation and the Navier–Stokes equation when the pressure gradient vanishes and $Pr = 1$. Other extended analogies which are valid for arbitrary Pr and turbulent flows, including the effects of the near-wall regions, are the modified Reynolds analogy, the Chilton–Colbourn analogy, the Prandtl analogy, and the Von–Karman Analogy.
4. **Electrodynamics-vorticity dynamics analogy** is constructed based on the idea that the magnetic field and the vorticity vector are both divergence-free. So, the electric current, the magnetic field, and the electric permeability are analogous to the circulation, the velocity, and the density, respectively. So, a similar relation inspired by the well-known Biot–Savart relation can be derived in vorticity dynamics.
5. **Free-surface-compressible flow analogy:** The flow of compressible fluid is similar to the creation of surface waves in free-surface flows. The Mach number, the pressure waves, and the shock wave are analogous to the Froude number, the surface waves, and the hydraulic jump.
6. **Other simple analogies:** In primary heat transfer and fluid mechanics courses, we teach students how to construct electrical circuit counterparts of piping networks and thermal systems by defining hydraulic or thermal resistances, respectively.

Effects: Thermo-fluid sciences owe their progress to well-known laws describing challenging effects. There are lots of such items. Here, we will review some of the celebrated effects related to thermo-fluid sciences.

1. The Kussner effect about the time-dependent lift force on an airfoil due to a cross-flow of gust.
2. The Kaye effect about the interaction of a jet of liquid with the free-surface of a complex fluid.

3. The fluid thread breakup about the decomposition of a specific mass of fluid into smaller parts, like what happens in the atomization of jets.

4. The Taylor column is a result of the Coriolis acceleration when a rotating fluid subjected to a solid body forms vertical Taylor columns.

5. The Stokes drift approximates the Lagrangian velocity of particles of free-surface flows in applications such as ocean flows.

6. The Langmuir circulation describes the appearance of counter-rotating vortices near the surface of a wind-driven water flow.

7. The Ekman drift discusses the spiral structure of the wind-driven flows under the effect of the Coriolis acceleration, especially near the earth's poles.

8. The Darwin drift is about the motion of a body inside a stationary fluid leading to the permanent displacement of the fluid particles.

9. The added/virtual mass results from the accelerating motion of a solid body inside an inviscid fluid.

10. The wave of translation related to the motion of solitary waves (solitons) on the free-surfaces.

11. The Coanda effect is a source of lift generation due to the creation of a low-pressure zone over the surface of a wall adjacent to a jet of fluid.

12. The Venturi effect is related to the pressure change in the Venturi tube.

13. The Torricelli effect speaks about the velocity of water leaving a tank at a depth of h, $\sqrt{2gh}$.

14. The Magnus effect is the appearance of a lift force due to the rotation of bluff bodies.

15. The Richardson's annular effect is the appearance of a velocity overshoot in the Stokes' second problem[2] when the free-stream oscillates.

16. Acoustic streaming is a second-order effect corresponding to the appearance of a non-zero mean advection term due to acoustic excitations in applications like an oscillating cylinder in a viscous flow [80].

NOTES

1. The second law is an indicator of the arrow of time.

2. The Stokes' first problem or Rayleigh's problem describes the motion of a fluid near an oscillating plate.

6 Complementary Topics

6.1 LINEARIZED THERMOELASTICITY

To avoid confusion in notation, in this section ε_{ij} and τ_{ij} denote the strain tensor and the Cauchy's stress tensor, respectively and e is the permutation symbol (Q6.1).

Conservation laws: The equation of motion is

$$\frac{\partial \tau_{ji}}{\partial x_j} + F_i = \rho \frac{\partial^2 d_i}{\partial t^2} \tag{6.1}$$

where F_i is the body force per unit volume and d_i is the displacement vector. The right-hand side is zero in the elasto-static equilibrium condition. The second Cauchy's equation or the conservation of angular momentum implies the symmetry of the stress tensor. As a result, we only need six constitutive relations for six unknown stress components.

Constitutive relations: The constitutive equation as a relation between stress and strain for a Hookian isotropic homogeneous solid is

$$\tau_{ij} = M_{ijkl}\varepsilon_{kl} = \lambda \varepsilon_{kk}\delta_{ij} + 2\mu\varepsilon_{ij} \tag{6.2}$$

where μ is the shear modulus, λ is the Lame's first constant, and M_{ijkl} is the elasticity tensor. The right-hand side can be obtained using a procedure similar to what we did for fluids. Due to the linearity of the constitutive equations, the strain–stress relation can be easily obtained after deriving ε_{kk} by replacing $i = j = k$ and using the identity $\delta_{kk} = 3$,

$$\varepsilon_{ij} = \frac{1}{2\mu}\left(\tau_{ij} - \frac{\lambda}{3\lambda + 2\mu}\delta_{ij}\tau_{kk}\right) = \frac{1+v}{E}\tau_{ij} - \frac{v}{E}\tau_{kk}\delta_{ij} \tag{6.3}$$

where E is the Young's modulus of elasticity and v is the Poisson's constant.

The linearized strain-displacement relation: The kinematic relation between an infinitesimal displacement and the strain tensor is

$$\varepsilon_{ij} = \frac{1}{2}\left(\frac{\partial d_i}{\partial x_j} + \frac{\partial d_j}{\partial x_i}\right) \rightarrow \tau_{ij} = \lambda \delta_{ij}\frac{\partial d_k}{\partial x_k} + \mu\left(\frac{\partial d_i}{\partial x_j} + \frac{\partial d_j}{\partial x_i}\right) \tag{6.4}$$

In solids, the deformations are not necessarily infinitesimal. In this context, we just present relations for the linearized elasticity.

Compatibility condition: Due to the symmetry of the strain tensor, only six components exist for the strain. Whilst, the displacement is a vector including three components. So, giving a known displacement field, it is easy to find the corresponding strain tensor by simple partial differentiations. But, the inverse is not possible, since we have six equations with three unknowns. So, we need some extra constraints known as compatibility conditions to perform integration for an over-determined system of equations. So, any set of six scalars cannot form a strain tensor.

DOI: 10.1201/9781032719405-6

We have to present a criterion to derive a strain tensor from a displacement field. The compatibility conditions guarantee the existence of a continuous, single-valued displacement field that satisfies Equation (6.4). Taking the second derivative of Equation (6.4) using the interchangeability (symmetry) of the mixed derivatives when the input function is continuous (Schwarz's theorem),

$$\frac{\partial^2 \varepsilon_{ij}}{\partial x_k \partial x_l} = \frac{1}{2}\left(\frac{\partial^3 d_i}{\partial x_k \partial x_l \partial x_j} + \frac{\partial^3 d_j}{\partial x_k \partial x_l \partial x_i}\right)$$

$$\frac{\partial^2 \varepsilon_{kl}}{\partial x_i \partial x_j} = \frac{1}{2}\left(\frac{\partial^3 d_k}{\partial x_i \partial x_j \partial x_l} + \frac{\partial^3 d_l}{\partial x_i \partial x_j \partial x_k}\right)$$

$$\frac{\partial^2 \varepsilon_{jl}}{\partial x_i \partial x_k} = \frac{1}{2}\left(\frac{\partial^3 d_j}{\partial x_i \partial x_k \partial x_l} + \frac{\partial^3 d_l}{\partial x_i \partial x_k \partial x_j}\right)$$

$$\frac{\partial^2 \varepsilon_{ik}}{\partial x_j \partial x_l} = \frac{1}{2}\left(\frac{\partial^3 d_i}{\partial x_j \partial x_l \partial x_k} + \frac{\partial^3 d_k}{\partial x_j \partial x_l \partial x_i}\right) \tag{6.5}$$

Then, the compatibility condition in tensor notation is

$$\frac{\partial^2 \varepsilon_{ij}}{\partial x_k \partial x_l} + \frac{\partial^2 \varepsilon_{kl}}{\partial x_i \partial x_j} - \frac{\partial^2 \varepsilon_{jl}}{\partial x_i \partial x_k} - \frac{\partial^2 \varepsilon_{ik}}{\partial x_j \partial x_l} = 0 \tag{6.6}$$

The derived compatibility relation can be expanded in $3^4 = 81$ equations. Only 6 of them are meaningful which can be written in the form

$$R_{il} = e_{ijk}e_{lmn}\frac{\partial^2 \varepsilon_{jm}}{\partial x_k \partial x_n} = 0$$

$$\mathbf{R} = \nabla \times (\nabla \times \boldsymbol{\varepsilon}) = \mathbf{0} \tag{6.7}$$

where $\mathbf{R}$ is the divergenceless second-order incompatibility tensor. Consider that the curl of a second-order tensor is a scalar, and the curl of a scalar is a second-order tensor. These six components can be written in two groups. For example, one of the equations in the first group is obtained by replacing $i = j = 1$ and $k = l = 2$

$$\frac{\partial^2 \varepsilon_{11}}{\partial x_2^2} + \frac{\partial^2 \varepsilon_{22}}{\partial x_1^2} - 2\frac{\partial^2 \varepsilon_{12}}{\partial x_1 \partial x_2} = 0 \tag{6.8}$$

and the case with $j = 1$, $i = k = 2$, and $l = 3$ corresponds to one of equations in the second group,

$$\frac{\partial^2 \varepsilon_{12}}{\partial x_2 \partial x_3} + \frac{\partial^2 \varepsilon_{23}}{\partial x_1 \partial x_2} - \frac{\partial^2 \varepsilon_{13}}{\partial x_2 \partial x_2} - \frac{\partial^2 \varepsilon_{22}}{\partial x_1 \partial x_3} = 0 \tag{6.9}$$

The Navier's equation: By substituting the stress–strain relation in the equilibrium equation and replacing the strains using the strain-displacement relations, the Navier's equation is obtained,

$$\frac{\partial \tau_{ji}}{\partial x_j} = \lambda\frac{\partial^2 d_k}{\partial x_k \partial x_i} + \mu\left(\frac{\partial^2 d_i}{\partial x_j \partial x_j} + \frac{\partial^2 d_j}{\partial x_i \partial x_j}\right) \rightarrow$$

$$\mu\frac{\partial^2 d_i}{\partial x_j \partial x_j} + (\lambda + \mu)\frac{\partial^2 d_j}{\partial x_j \partial x_i} + F_i = 0 \tag{6.10}$$

The Biharmonic equation: Taking the divergence of both sides of the Navier's equation in elastostatic mode, and assuming the body forces to be divergence-free, using the Schwarz' theorem we have

$$\frac{\partial^3 d_j}{\partial x_i \partial x_i \partial x_j} = 0 \tag{6.11}$$

This time, take the Laplacian of both sides of the Navier's equation and neglect the body forces to obtain the biharmonic equation

$$\frac{\partial^4 d_i}{\partial x_k \partial x_k \partial x_m \partial x_m} = 0 \tag{6.12}$$

Introducing the Airy displacement potential function ϕ similar to the potential function concept in fluid mechanics, in the case of plain strain,

$$\begin{aligned}
\tau_{11} &= \frac{\partial^2 \phi}{\partial x_2^2} \\
\tau_{22} &= \frac{\partial^2 \phi}{\partial x_1^2} \\
\tau_{12} &= -\frac{\partial^2 \phi}{\partial x_1 \partial x_2}
\end{aligned} \tag{6.13}$$

along with the equilibrium condition

$$\nabla^2 (\tau_{11} + \tau_{22}) = 0 \rightarrow \nabla^2 \nabla^2 \phi = \nabla^4 \phi = 0 \tag{6.14}$$

where ∇^4 is the bilaplacian operator and the final equation is the biharmonic equation that also governs the motion of creeping flows.

The Lame's equation is the elastodynamic version of Navier's equation

$$\mu \frac{\partial^2 d_i}{\partial x_j \partial x_j} + (\lambda + \mu) \frac{\partial^2 d_j}{\partial x_i \partial x_j} + F_i = \rho \frac{\partial^2 d_i}{\partial t^2} \tag{6.15}$$

The Clausius–Duhem inequality, the second law for solids: The specific entropy of solids is usually shown by η. Using the identity (5.62), the Clausius–Duhem inequality is obtained,

$$\rho \dot{\eta} \geq \frac{\rho Q}{T} - \nabla \cdot \left(\frac{\mathbf{q}}{T} \right) \rightarrow \rho T \dot{\eta} - \rho Q + \nabla \cdot \mathbf{q} - \frac{\mathbf{q} \cdot \nabla T}{T} \geq 0 \tag{6.16}$$

where Q is the source of the heat per unit mass, the dotted symbols denote time-derivatives, and $D = \rho T \dot{\eta} - \rho Q + \nabla \cdot \mathbf{q}$ is the internal dissipation. Using the energy equation,

$$\rho \dot{e} = \rho Q - \nabla \cdot \mathbf{q} + \boldsymbol{\tau} : \nabla \mathbf{u} \rightarrow \nabla \cdot \mathbf{q} = -\rho \dot{e} + \rho Q + \boldsymbol{\tau} : \nabla \mathbf{u} \tag{6.17}$$

and substituting the divergence of heat flux vector from the energy equation in the Clausius–Duhem inequality, we obtain the following version of the second law,

$$-\frac{\mathbf{q} \cdot \nabla T}{T} \geq \rho (\dot{e} - T \dot{\eta}) - \boldsymbol{\tau} : \nabla \mathbf{u} \tag{6.18}$$

If the process is isentropic, then $\dot{\eta}$ should become zero. Dissipation function with the unit of power per unit volume is defined as

$$\Phi \equiv -\frac{\mathbf{q} \cdot \nabla T}{T} - \rho(\dot{e} - T\dot{\eta}) + \boldsymbol{\tau} : \nabla \mathbf{u} \geq 0 \qquad (6.19)$$

This is called the dissipation inequality. The first term represents thermal dissipation which is positive when the heat flux vector and the temperature gradient are of opposite signs.

If we ignore the thermal dissipation such as what occurs in rapid or adiabatic processes, the new equation is often called the Clausius–Plank inequality. Consider a reversible process with $Q = 0$, using the energy equation, it is easy to show that the Clausius–Plank relation is

$$T\dot{\eta} = \dot{e} - \frac{1}{\rho}\boldsymbol{\tau} : \nabla \mathbf{u} \rightarrow \rho\dot{\eta} = -\frac{\nabla \cdot \mathbf{q}}{T} \qquad (6.20)$$

Conduction in solids: We may easily obtain the energy equation for solids by replacing $\mathbf{u} = \mathbf{0}$ in the energy equation for fluids. However, an alternative strategy for deriving the conduction equation in solids [92] is based on the definition of the internal dissipation for reversible processes and Equation (6.16),

$$D = \rho T\dot{\eta} - \rho Q + \nabla \cdot \mathbf{q} = 0 \qquad (6.21)$$

The constitutive relations of heat flux and entropy are,

$$q_i = -k\frac{\partial T}{\partial x_i}$$
$$\rho\eta = \frac{C}{T_0}T \qquad (6.22)$$

where $C = \rho c_v$ is the volumetric heat capacity,

$$C\frac{T}{T_0}\dot{T} - \rho Q - k\frac{\partial^2 T}{\partial x_i \partial x_i} = 0 \qquad (6.23)$$

By assuming $T = T_0$ in the time-derivative term, which is a common presumption in heat conduction of solids [92], the linear governing equation of heat conduction in solids becomes,

$$C\dot{T} = \nabla \cdot (\mathbf{k} \cdot \nabla T) + \rho Q \qquad (6.24)$$

Thermoelasticity: Thermal scientists may be interested in computing the stress field in a solid due to non-uniform temperature distribution. The thermoelastic constitutive equation for a Hookian solid is

$$\tau_{ij} = M_{ijkl}\varepsilon_{kl} - \alpha(T - T_0)\delta_{kl} = \lambda\varepsilon_{kk}\delta_{ij} + 2\mu\varepsilon_{ij} - \alpha\delta_{ij}(T - T_0) \qquad (6.25)$$

where T_0 is the reference temperature, $T - T_0$ is the excess temperature difference, $\alpha = \frac{\beta E}{1-2v}$, v is the Poisson's constant, β is the thermal expansion coefficient, E is the Young's Modulus of elasticity, and M_{ijkl} is the elasticity tensor. The strain–stress equation is

$$\varepsilon_{ij} = \frac{1+v}{E}\tau_{ij} - \frac{v}{E}\tau_{kk}\delta_{ij} + \beta(T - T_0)\delta_{ij} \tag{6.26}$$

As a good exercise, derive the compatibility conditions for thermoelastic problems. The relation between the displacement vector and the strain tensor called the kinematic condition is

$$\varepsilon_{ij} = \frac{1}{2}\left(\frac{\partial d_i}{\partial x_j} + \frac{\partial d_j}{\partial x_i}\right)$$

By substituting the stress–strain relation in the equilibrium equation and replacing the strains using the strain-displacement equation, Navier's equation is obtained,

$$\mu\frac{\partial^2 d_i}{\partial x_l \partial x_l} + (\lambda + \mu)\frac{\partial^2 d_l}{\partial x_l \partial x_i} + F_i - \alpha\frac{\partial T}{\partial x_i} = \rho\frac{\partial^2 d_i}{\partial t^2} \tag{6.27}$$

In uncoupled thermoelasticity, the stress terms in the energy equation are omitted and the temperature field is solved at the first stage without the need for the stress field. Then, the displacement equations are solved based on the computed temperature field. In coupled thermoelasticity, equations of displacement and temperature have to be solved simultaneously in a loop. The first and second Danilovskaya problems and the Sternberg and Chakravorty problem are classical cases in coupled thermoelasticity regarding thermal shocks.

Introducing the Airy displacement potential function ϕ in the plain strain equation in cartesian coordinates along with the equilibrium condition, the non-homogenous form of the biharmonic equation reads

$$\nabla^2(\tau_{11} + \tau_{22}) = \nabla^4\phi = \frac{\alpha}{1-v}\nabla^2 T \tag{6.28}$$

Analogies with fluid mechanics: Comparing the equations presented for elastic solids with those of fluids, we obtain some analogies (Q6.2).

1. Similar to the first and second coefficients of viscosity in fluid mechanics, in elasticity theory, we have the following constants two of which are independent for an isotropic medium.

   a. The Young's modulus (E): It is an indicator of the tensile strength of the material, $E = 3K(1 - 2v)$.
   b. The shear modulus or Lame's second coefficient (μ): It is an indicator of the strength of material against shear (at constant volume) similar to the concept of dynamic viscosity of fluids $\mu = \frac{E}{2(1+v)} = \frac{3K(1-2v)}{2(1+v)}$.
   c. The bulk modulus (K): It is an indicator of the strength of material against volume change and is proportional to the inverse of compressibility $K = \lambda + \frac{2}{3}\mu$. It is the counterpart of Stokesity in fluids.
   d. The Lame's first constant $\lambda = \frac{2\mu v}{1-2v} = K - \frac{2}{3}\mu$ is the counterpart of the second coefficient of viscosity in fluids.

     e. The Poisson's constant (v) is defined as the amount of transversal expansion divided by the amount of axial compression, $v = \frac{E}{2\mu} - 1 = \frac{3K-2\mu}{2(3K+\mu)}$.

2. Based on the Stokes hypothesis, the Stokesity, and Young's modulus are zero for fluids. $K = 0$ is equivalent to infinite density, based on the definition of the bulk modulus.

3. In fluid mechanics, accepting the Stokes hypothesis means that $v = -1$, $K = E = 0$. However, Poisson's constant can be defined in fluids as a function of density. Kozachok (2019) [10] declared that values of $v = -1, 0.5, 0$ correspond to $K = 0, \infty, \frac{2}{3}\mu$, respectively.

4. Instead of strains, the strain-rates or the velocity of strains is important in fluids. A flow is a continuous deformation in time. It means that each particle of fluid continually changes its position with time. Therefore, the displacement is a vague parameter and the strain-rate has to be computed. In fluids and viscoelastic materials, the deformation is not necessarily a result of strain. For instance, particles of a fluid in a static condition have Brownian displacement without deformation. But, if a force is exerted on a solid, it will deform and the particles will displace. However, the exerted force changes the velocity of fluid particles along their pathline. The strain and the strain-rate tensors are related to the displacement and velocity, respectively in fluids and solids. The dependent variable in the Navier–Stokes equation and the Lame's equation are velocity and displacement, respectively. In fluids, the displacement itself does not contribute to the stress tensor. If the strain exists but does not vary with time, the forces obtained from the stress tensor are zero.

5. In fluids, the acceleration should be computed using the material derivative of the velocity vector instead of simple differentiation with respect to time. This is a source of nonlinearity in the complex motion of fluids.

6. The main difference between Lame's equation and the Navier–Stokes equation is the pressure term in the Navier–Stokes equation. So, the continuity equation has to be added to the system of governing equations to close them. The appearance of pressure just in the momentum equation is the source of many problems in the solution of the equations of motion for incompressible flows. In order to deal with this problem there are many strategies,

    • Simply omitting the pressure term from the Navier–Stokes equation which yields the Burgers equation, Equation (4.80).

    • Obtaining an explicit Poisson-type equation for pressure (Section 4.1.2.2).

    • Eliminating the pressure term from the momentum equation by taking the curl of the Navier–Stokes equation (Chapter 7).

    • Using the artificial compressibility term in numerical methods.

    • Using the weakly compressible form of the equations (Section 4.1.2.3).

    • Deriving an alternative form of equations such as what we did for shallow water open-channel flows (Section 4.1.2.6).

7. Half of the vorticity vector (the angular velocity vector) in fluid mechanics is similar to the rotation vector in solid mechanics.

8. A compatibility condition may be obtained for fluids exactly similar to the solids. But, the strain tensor and the displacement vector should be replaced by the strain-rate tensor and the velocity vector, respectively.

9. Assumptions such as infinitesimal or large deformation are meaningless for fluids. All deformations of fluids are small due to weak intermolecular bonds.

6.2 FUNDAMENTALS OF ELECTRODYNAMICS

In various applications such as elctrocapillary, electrowetting,[1] magnetic pumps, piezomateials, thermal design of transformers, nanoscale transistors, electrophoresis, electroosmosis, plasma flows, photovoltaic cells, direct energy conversion, magnetohydrodynamics, high-voltage circuit breakers, ferrofluids, non-destructive testing involving eddy current, thermal radiation, and measurement devices, we are engaged with effects of a magnetic or an electric field.

The electric and magnetic fields deal with a stationary or a moving charge, respectively. Here, we have ignored the small-scale effects encountered in fluctuating, stochastic, and quantum electrodynamics,[2] and only fundamentals of classical electrodynamics are discussed. From a historical viewpoint, Faraday in 1820 and Maxwell in 1873 presented experimental and theoretical investigations of electromagnetism, respectively. Different governing equations of electrodynamics consist of constitutive relations and fundamental relations such as the Lorentz force and Maxwell's equations (Jackson 1999) [99].

Constitutive relations: Similar to other branches of continuum mechanics, there are constitutive relations in electrodynamics, which can be violated under certain conditions,

$$
\begin{aligned}
\mathbf{D} &= \varepsilon \mathbf{E} \\
\mathbf{B} &= \mu \mathbf{H} \\
\mathbf{J} &= \sigma[\mathbf{E} + \mathbf{v} \times \mathbf{B}]
\end{aligned}
\tag{6.29}
$$

where $\mathbf{D}, \mathbf{E}, \mathbf{B}, \mathbf{H}, \mathbf{v}, \varepsilon, \sigma, \mu$ are the displacement vector, the electric field, the magnetic field, the magnetizing field, the velocity of moving charges, the permittivity, the conductance, and the permeability, respectively. The third equation is the general form of the well-known Ohm's law.

The Lorentz force: The Lorentz force is one of four fundamental forces in physics.[3] It can be proved that the electromagnetic field exerts the following Lorentz force on a charged particle,

$$
\mathbf{f}_L = q\mathbf{E} + q\mathbf{v} \times \mathbf{B}
\tag{6.30}
$$

As mentioned in previous chapters, it is the density of forces that enters the conservation laws. So, we need to compute the Lorentz force per unit volume

$$
\mathbf{F}_L = \rho_c \mathbf{E} + \mathbf{J} \times \mathbf{B}
\tag{6.31}
$$

where ρ_c and $\mathbf{J}$ are the electric charge density and the electric current density, respectively. Integration of ρ and $\mathbf{J}$ over volume and surface, respectively, yields the total charge inside the volume and the current flux crossing the area.

6.2.1 MAXWELL'S EQUATIONS

The Maxwell's equations consist of four partial differential equations.

The Gauss law for electric field is a scalar equation and presents a relation for the divergence of the electric field,

$$\nabla \cdot \mathbf{E} = \frac{\rho_c}{\varepsilon_0} \qquad (6.32)$$

Defining $\mathbf{E} = -\nabla\phi$ for the electrostatic case where ϕ is the electric potential, yields

$$\nabla^2 \phi = -\frac{\rho}{\varepsilon_0} \qquad (6.33)$$

which is a Poisson-type equation. For vacuum, the right-hand side is zero, and Equation (6.33) reduces to the Laplace equation. It is simple to show that based on the definition of the electric potential, the Lorentz density force equals $-\varepsilon\nabla^2\phi\mathbf{E}$. The displacement field is defined as

$$\nabla \cdot \mathbf{D} = \rho_f \qquad (6.34)$$

in which ρ_f is the free-charge density.

The Gauss law for magnetism is a scalar equation and indicates that the magnetic field is divergence-free,

$$\nabla \cdot \mathbf{B} = 0 \qquad (6.35)$$

Maxwell–Faraday law or the Faraday's law of induction is a vector equation and presents a relation for the curl of the electric field as a function of temporal variation of the magnetic field. So, a varying magnetic field can induce an electric field,

$$\nabla \times \mathbf{E} = -\frac{\partial \mathbf{B}}{\partial t} \qquad (6.36)$$

It is obvious that for a zero or steady-state magnetic field, the electric field is curl-free. So, we are allowed to define the electric field as a gradient of a scalar field, called the electric potential. Taking the curl of both sides of Equation (6.36),

$$\mu_0\varepsilon_0 \frac{\partial^2 \mathbf{E}}{\partial t^2} - \nabla^2\mathbf{E} = 0 \qquad (6.37)$$

where $\mu_0\varepsilon_0 = \frac{1}{c^2}$ and c is the velocity of light. The result of a varying magnetic field is the creation of the eddy or Foucault's currents. The name "eddy" has been originated from the similarity of the eddy currents with the eddies in turbulent flows. Another consequence of the Maxwell–Faraday's law is Lenze's law which is out of the scope of this book.

The Ampere's law is a vector equation and presents a relation for the curl of the magnetic field as a function of the temporal derivative of the electric field. So, a varying electric field can induce a magnetic field,

$$\nabla \times \mathbf{B} = \mu_0 \left(\mathbf{J} + \varepsilon_0 \frac{\partial \mathbf{E}}{\partial t} \right) \tag{6.38}$$

Taking curl of both sides of Equation (6.38),

$$\mu_0 \varepsilon_0 \frac{\partial^2 \mathbf{B}}{\partial t^2} - \nabla^2 \mathbf{B} = 0 \tag{6.39}$$

Using the constitutive relation $\mathbf{B} = \mu \mathbf{H}$

$$\nabla \times \mathbf{H} = \mathbf{J}_f + \frac{\partial \mathbf{D}}{\partial t} \tag{6.40}$$

where $\mathbf{J}_f$ is the free-current density. Pay attention that Maxwell's equations consist of two vector equations and two scalar relations. But, we only have two unknown vectors. So, Maxwell's equations are over-determined. It should be noted that Gauss's laws are just constraints, and should be satisfied during the solution process.

6.2.2 CONNECTIONS WITH MECHANICAL SCIENCES

Analogies with fluid mechanics: Similarities between the governing equations of electrodynamics and fluid mechanics motivate us to find similarities between these two sophisticated sciences. The celebrated Biot–Savart law is a bridge between electrodynamics and vorticity dynamics,

$$\mathbf{B} = \frac{\mu_0 I}{4\pi} \int_{wire} \frac{\mathbf{dl} \times \hat{\mathbf{r}}}{\mathbf{r}^2} \tag{6.41}$$

Here, $\mathbf{B}$ is the induced magnetic field from a current in a wire with a line element $\mathbf{dl}$ along it. The Biot–Savart law in vorticity dynamics speaks about the induced velocity by the circulation of a vortex tube. Hence, vortex tube, circulation, velocity, and vorticity are analogous to the wire, permeability, magnetic field, and electrical current, respectively.

Other common equations are the conservation of mass and charge

$$\frac{\partial \rho}{\partial t} + \nabla \cdot \mathbf{J} = 0$$

$$\frac{\partial \rho_f}{\partial t} + \nabla \cdot (\rho_f \mathbf{u}) = 0 \tag{6.42}$$

where ρ_f is the density of the fluid. Other similarities can be found in literature such as the similarity of sound velocity, and light velocity, compressibility factor and permittivity, virtual mass and relativistic mass [100].

Magneto–hydro-dynamics (MHD) involves applications that an electrically conducting fluid like plasma or liquid metal is exposed to a magnetic field. The

MHD generators are used to directly convert thermal and kinetic energies into electricity. The important dimensionless number in MHD flow is the Hartman number $Ha = BL\sqrt{\frac{\sigma}{\mu}}$, which is the ratio of the electromagnetic force to the viscous force. The governing equations of MHD flows include the conservation of mass, the Navier–Stokes equations with the Lorentz force as a source term ($\mathbf{J} \times \mathbf{B}$), the Maxwell–Faraday, the Ampere, and the Ohm's laws.

Electro–hydro–dynamics (EHD) investigates the motion of electrically charged or ionized fluids exposed to an electric field. Since the Lorentz force in the EHD context is weak, the EHD applications are mostly involved with nanoscale geometries. Two important applications of EHD are electrophoresis and electroosmosis, which respectively are the study of the motion of particles in an electric fields and the motion of liquids across a membrane or a porous material under the action of an electric potential.

Again, the Lorentz force with zero magnetic field should be added to the Navier–Stokes equation as a source term. There are important phenomena and concepts in EHD flows such as the electric-double layer, the Debay length, the zeta potential, and the mobility that are out of the scope of this book.

Joule heating: When an electric current passes across a material, the electrons start to align with the applied electric field. On the other hand, due to the interaction of electrons and ions in the structure of the conductor as a random scattering process, electrical resistivity appears. As a result of the mentioned scattering, the kinetic energy of electrons is converted to the thermal energy called the Joule/Ohmic/resistive heating.

Joule heating appears as a volumetric source term in the energy equation,

$$Q = \frac{Power}{\forall} = \mathbf{J} \cdot \mathbf{E} \qquad (6.43)$$

Using the constitutive relation $\mathbf{J} = \sigma \mathbf{E}$,

$$Q = \frac{\mathbf{J} \cdot \mathbf{J}}{\sigma} = RJ^2 \qquad (6.44)$$

Joule heating arises in applications such as heating electric elements, coils, light filaments, transistors, photovoltaic cells, metal-oxide-semiconductor devices, and electrical wires. It should be noted that besides the Joule heating, there are other mechanisms of dissipation of energy in electrodynamical applications, namely the non-radiative electron-hole generation and recombination [101–103] (Q6.3).

Piezoelectric effect: If an external mechanical stress is applied to a piezoelectric material,[4] an internal electric charge will be induced in the matter. Piezogenerators are electromechanical devices to produce electricity from unused dynamic vibrations in energy harvesting applications. Also, it is possible to utilize a piezomaterial to create mechanical displacement by applying electrical energy [104]. This phenomenon is called the inverse piezoelectric effect. Governing equations of the piezoelectric effect are

1. Maxwell's equations relate the electrical field to the scalar electric potential. Also, the displacement file is divergenceless (3+1 equations).

$$D_{i,i} = 0, \quad E_i = -\phi_{,i} \tag{6.45}$$

2. The constitutive electrical relations (3 equations) in two versions

$$
\begin{aligned}
D_i &= \xi_{ij}E_j + \chi_{ijk}\varepsilon_{jk} \\
D_i &= \xi_{ij}E_j + d_{ijk}\tau^c_{jk}
\end{aligned}
\tag{6.46}
$$

which presents a linear relationship between the electrical displacement and the electric field plus the piezoelectric part. ξ_{ij} is the symmetric dielectric permittivity tensor. χ_{ijk} and d_{ijk} are the third-rank piezoelectric-strain tensors, which map the symmetric strain or stress tensor to the electrical displacement vector.

3. The constitutive mechanical relations (6 equations) in two versions

$$
\begin{aligned}
\tau^c_{ij} &= E_{ijkl}\varepsilon_{kl} - \chi_{kij}E_k \\
\varepsilon_{ij} &= F_{ijkl}\tau^c_{kl} + d_{kij}E_k
\end{aligned}
\tag{6.47}
$$

which present a linear relationship between the stress and strain tensors (Hook's law) plus the piezoelectric part. E_{ijkl} and F_{ijkl} are the elasticity and the compliance tensors, respectively. χ_{kij} is the converse piezoelectric third-order tensor, which maps the electrical field to the stress tensor.

4. Equations of mechanical equilibrium (3 equations)

$$\tau^c_{ij,j} + F_i = \rho\frac{\partial^2 u_j}{\partial t^2} \tag{6.48}$$

5. The linear displacement-strain relations (6 equations)

$$\varepsilon_{ij} = \frac{1}{2}(u_{i,j} + u_{j,i}) \tag{6.49}$$

The set of governing equations consists of 22 equations and 22 unknowns, including 3 velocities, 6 strain components, 6 stress components, 3 electric field components, 3 displacement components, and ϕ [105]. Also, there are three subcases of piezomaterials called pyroelectrics, ferroelectrics, and flexoelectrics. Pyroelectric material is a type of piezomaterial, which can generate electricity from time fluctuation of temperature. The ferroelectric material is a subset of pyroelectric materials, which can produce electrical energy using reversible polarization. The flexoelectric materials not only harvest energy from the stress tensor (like the piezoelectric material) but also generate electricity from the gradient of strain.

Thermoelectric effect is the reversible generation of electricity from a spatial temperature difference between cold and hot reservoirs. The process of direct generation of electricity (as a heat engine) when the junction of two non-identical materials is heated is called the Seebeck effect. The inverse process that is the generation of heat via temperature difference when a voltage is applied to the junction of two non-identical materials is the Peltier effect. The Peltier effect can be used to construct thermoelectric refrigerators.

The Thomson effect is the continuous extension of the Peltier effect, which describes the reversible heating or cooling of a single conductor subjected to a temperature gradient when a current is sent through the conductor. Heat can only be generated during Joule heating due to its irreversible nature. But, the Thomson effect may lead to both heating or cooling of the conductor. The Thomson effect appears along with the Seebeck effect when the thermoelectric property of the matter is a function of temperature.

The governing equations to predict the behavior of an isotropic thermoelectric material are

1. The constitutive relation for the current density (the extended Ohm's law)

$$\mathbf{J} = -\sigma \left(\nabla \phi + \alpha \nabla T \right) \tag{6.50}$$

or the electric field-based version

$$\mathbf{E} = \frac{\mathbf{J}}{\sigma} + \alpha \nabla T \tag{6.51}$$

where σ and α are the electrical conductivity and the Seebeck coefficient, respectively. The last term on the right-hand side is the electromotive field that is a non-electrical source of electricity. Equation (6.50) implies that the electric current can be generated due to the electrical conductivity term (the first term on the right-hand side) and the Seebeck effect (the second term in the right-hand side).

2. The constitutive relation for the heat flow density (the extended Fourier's law)

$$\mathbf{q} = -k\nabla T + \alpha T \mathbf{J} \tag{6.52}$$

3. The current continuity relation

$$\nabla \cdot \mathbf{J} = 0 \tag{6.53}$$

4. The conservation of energy

$$-\nabla \cdot \mathbf{q} + Q_{Joule} = \rho c_p \frac{\partial T}{\partial t} \tag{6.54}$$

where the source term denotes the volumetric Joule heating $Q_{Joule} = \mathbf{J} \cdot \mathbf{E} = -\mathbf{J} \cdot \nabla \phi = \frac{J^2}{\sigma} + \mathbf{J} \cdot \alpha \nabla T$. Substituting two constitutive relations and the Joule heating term in the energy equation, the modified energy equation containing the thermoelectric terms reads

$$\rho c_p \frac{\partial T}{\partial t} = \nabla \cdot (k\nabla T) - \nabla \cdot (\alpha T \mathbf{J}) + \mathbf{E} \cdot \mathbf{J} \tag{6.55}$$

The first term in the right-hand side represents thermal conduction, the second term $(-\nabla \cdot (\alpha T \mathbf{J}))$ encounters the Peltier effect and Thomson effect when the Seebeck coefficient is a function of temperature $(\alpha(T))$, and the third term $(\mathbf{E}.\mathbf{J})$ is the Joule heating source term.

To explicitly demonstrate the Peltier effect, the second term plus the Joule heating term can be rewritten using the identities for the divergence of product and gradient of product, and the fact that $\nabla \cdot \mathbf{J} = 0$,

$$
\begin{aligned}
& - \quad \nabla \cdot (\alpha T \mathbf{J}) + \mathbf{E}.\mathbf{J} \\
&= \quad -(\alpha T)\nabla \cdot \mathbf{J} - \mathbf{J} \cdot \nabla(\alpha T) + \left(\frac{\mathbf{J}}{\sigma} + \alpha \nabla T\right) \cdot \mathbf{J} \\
&= \quad 0 - \mathbf{J} \cdot (\alpha \nabla T + T \nabla \alpha) + \frac{J^2}{\sigma} + \alpha \nabla T \cdot \mathbf{J} \\
&= \quad -\mathbf{J} \cdot T \nabla \alpha + \frac{J^2}{\sigma} \\
&= \quad \kappa \mathbf{J} \cdot \nabla T + \frac{J^2}{\sigma} \quad\quad\quad\quad\quad\quad (6.56)
\end{aligned}
$$

The last relation is obtained using $\nabla \alpha = \frac{d\alpha}{dT}\nabla T$ and a new constant $\kappa = -T\frac{d\alpha}{dT}$ that is called the Thomson coefficient.

5. The entropy equation obtained in Equation (5.63) can be extended to include the thermoelectric and the Joule heating effects. The new field equation for entropy that neglects the dissipation term and other sources of heat generation is

$$
\begin{aligned}
\rho\frac{Ds}{Dt} &= \quad -\nabla \cdot \left(\frac{\mathbf{q}_{TE}}{T}\right) - \frac{1}{T^2}\mathbf{q}_{TE} \cdot \nabla T + \frac{\mathbf{E} \cdot \mathbf{J}}{T} \\
&= \quad \nabla \cdot \left(\frac{k\nabla T - \alpha T \mathbf{J}}{T}\right) - \frac{1}{T^2}(-k\nabla T + \alpha T \mathbf{J}) \cdot \nabla T + \frac{\frac{J^2}{\sigma} + \alpha \mathbf{J} \cdot \nabla T}{T} \\
&= \quad \nabla \cdot \left(\frac{k\nabla T}{T} - \alpha \mathbf{J}\right) + \frac{1}{T^2}\nabla T(k\nabla T) - \frac{\alpha}{T}\nabla T \cdot \mathbf{J} + \frac{J^2}{\sigma T} + \frac{\alpha}{T}\nabla T \cdot \mathbf{J} \\
&= \quad \nabla \cdot \left(\frac{k\nabla T}{T} - \alpha \mathbf{J}\right) + \frac{1}{T^2}\nabla T(k\nabla T) + \frac{J^2}{\sigma T} \quad\quad\quad (6.57)
\end{aligned}
$$

The new terms $-\alpha \mathbf{J}$ and $\frac{J^2}{\sigma T}$ are the reversible (Peltier) and the irreversible (Joule heating) sources of entropy production, respectively. The Peltier term is proportional to the first power of $\mathbf{J}$, and its sign may change when the current is reversed. The Joule heating term is always positive due to the second power of $\mathbf{J}$ [106].

6.3   FLOW IN POROUS MEDIA

Henry Darcy was the first who performed experiments on the flow of water through beds of sand acting like a porous medium. There are many applications

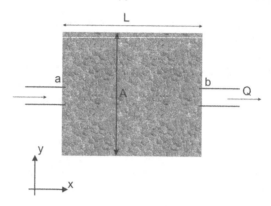

Figure 6.1 Flow through a porous reservoir.

in groundwater flows, earth sciences, the petroleum industry, biology, and energy-related engineering in which porous materials play a key role (Q6.4).

Here, we are to introduce a new constitutive relation for flow through porous media that can simulate complexities of flow inside randomly distributed pores of a porous medium [107]. Based on the introduced constitutive relation, the modified version of the momentum equation to model flows in porous media will be derived.

Consider a cavity with the length of L filled with a homogenous porous medium as shown in Figure 6.1. The inlet and the outlet ports are labeled with a and b, respectively. Cross-sectional area of the cavity is A and the volumetric flow rate is Q. Ignoring gravity, the relation for computing the flow rate is written as follows,

$$Q = -\frac{kA(p_b - p_a)}{\mu L} \tag{6.58}$$

where k is the permeability of the media (m^2), and the negative sign demonstrates that the direction of flow is along the positive x-direction. Equation (6.58) may be used to compute the intrinsic permeability of a single-phase flow through a porous medium. Dividing both sides of the equation by area, the vectorial form of the relation is obtained,

$$\mathbf{q} = -\frac{k}{\mu}\nabla p \tag{6.59}$$

where $\mathbf{q}$ is the flow discharge per unit area (m/s). To compute the velocity from the discharge vector, we define the porosity of the porous medium,

$$\mathbf{u} = \frac{\mathbf{q}}{\phi} \tag{6.60}$$

Equation (6.60) implies that only a fraction of the overall volume of the material is available for the fluid to stream. The main assumption behind Darcy's model is the low Reynolds number of flow between pores, which is equivalent to slow, laminar Stokes flow. The Reynolds number is defined based on the grain diameter $Re = \frac{U_c d}{\nu}$.

However, there are cases with coarse pores where the flow between the pores becomes turbulent even in small gaps. The Darcian-flow regime corresponds to the flow with the Reynolds number less than unity.

The Darcy's law: Consider the Navier–Stokes equation in creeping mode,

$$0 = -\frac{\partial p}{\partial x_i} + \rho g_i + F_{i,viscous} \tag{6.61}$$

where the viscous force has a linear relation with the velocity following the equation $F_{i,viscous} = (-k_{ij}^{-1}\mu\phi)u_j$. k_{ij} is the second-order permeability tensor. The off-diagonal elements are zero and the diagonal components are identical for an isotropic porous material $k_{ij} = k\delta_{ij}$. Multiplying by k_{ni} and dividing by $\mu\phi$

$$(-k_{ij}^{-1}\mu\phi)u_j = \frac{\partial p}{\partial x_i} - \rho g_i$$

$$k_{ni}k_{ij}^{-1}u_j = -\frac{k_{ni}}{\mu\phi}(\frac{\partial p}{\partial x_i} - \rho g_i) \tag{6.62}$$

and using the relation $k_{ni}k_{ij}^{-1} = \delta_{nj}$, it yields

$$u_n = -\frac{k_{ni}}{\mu\phi}(\frac{\partial p}{\partial x_i} - \rho g_i) \rightarrow$$

$$\mathbf{q} = -\frac{k}{\mu}(\nabla p - \rho\mathbf{g}) \tag{6.63}$$

The last equality holds just for isotropic porous media. If we suppose that the porous medium consists of spherical pores with an average diameter of d [108], then

$$k = \frac{d^2\varepsilon^3}{175(1-\varepsilon^2)} \tag{6.64}$$

The Darcy–Forchheimer law is one of the non-Darcian models which can account for internal effects at higher Reynolds numbers. In this mode, a new term called the Forchheimer inertia term is added to the equation to consider the nonlinear behavior of the flow in porous media,

$$\frac{\partial p}{\partial x} = -\frac{\mu}{k}q - \frac{\rho}{k_i}q^2 \tag{6.65}$$

where k_i is the inertial permeability.

The Klinkenberg effect: This phenomenon is seen in gaseous flows with small characteristic length-scales such as flows toward fine sand or the porous media containing nanostructures. Interaction of flow with the walls of pores imposes a new viscous force called Knudsen friction, which leads to the Knudsen diffusion. In this context, Darcy's model is modified by defining an effective permeability k_e

$$q = -\frac{k_e}{\mu}\frac{\partial p}{\partial x}$$

$$k_e = k\left(1 + \frac{b}{p}\right) \tag{6.66}$$

where b is the Klinkenberg parameter, which depends on the gas and the porous media structure [109].

Brinkman–Darcy law: This non-Darcian model is used when the grains inside the porous media are porous themselves. A new term added to the Darcy model is called the Brinkman term,

$$-\beta \nabla^2 \mathbf{q} + \mathbf{q} = -\frac{k}{\mu} \nabla p \tag{6.67}$$

where β is the effective viscosity [110].

Other techniques to model the behavior of flow in porous media are direct modeling [111], random models [112], reconstruction of the pore network [113], fractal models [114], and the Brinkman model with local volume-averaging [115]. Sometimes the flow over bluff bodies may be modeled as flow in porous media called the Brinkman penalization model, and the penalized version of the Navier–Stokes equations for porous media can be solved [116].

6.4   TRAFFIC FLOW: CONSERVATION OF VEHICLES

The traffic flow modeling intends to simulate the motion of cars in streets. When the number of cars increases enough, the characteristics of the system become deterministic, and we are allowed to use continuum mechanics to describe the system. Anyway, finding a mathematical model for such a nonlinear system involving human decisions is quite rigorous. However, we try to construct a physical model similar to that of compressible flows to predict the flow of cars.

The traffic density ρ is the number of vehicles per unit length of the road, the flow rate $Q = \rho v$ is defined as the number of vehicles passing a point per unit time, and v is the mean flow velocity. It should be noted that the continuum macroscopic models are only valid when the density of cars is large enough. Then, a definition similar to the Kn number can be presented called the traffic-Knudsen number based on the vehicle's mean-free-path and the characteristic size of the road. The high traffic-Kn limit denotes the free condition, the intermediate traffic-Kn corresponds to the stop-and-go condition, and very low traffic-Kn implies the stop condition. The final goal of the following models is to forecast the optimum design of traffic, concerning the fluent movement of vehicles without traffic congestion.

1. The kinematic first-order LWR model: Lighthill & Whitham [117] and Richards [118] presented the simplest conservation of vehicle equation called the Lighthill–Whitham–Richards model,

$$\frac{\partial \rho}{\partial t} + \frac{\partial Q}{\partial x} = 0 \tag{6.68}$$

This relation states that the time-rate of change of density is balanced by the spatial variation of the flow of cars.

2. The second-order version of the LWR model reads

$$\frac{\partial \rho}{\partial t} + A \frac{\partial \rho}{\partial x} + B \frac{\partial^2 \rho}{\partial t^2} = C \frac{\partial^2 \rho}{\partial x^2} \tag{6.69}$$

A is the speed of wave, *B* is the time-constant of the speed change, and *C* is the diffusion coefficient. The third and fourth terms are new modifications and correspond to the variation of the speed of vehicles (the inertial term) and the consciousness of drivers against changes in other locations (the diffusion term), respectively.

3. The Payne second-order model:

$$\frac{\partial v}{\partial t} + v\frac{\partial v}{\partial x} + \frac{c_0^2}{\rho}\frac{\partial \rho}{\partial x} = \frac{v_e - v}{B} \tag{6.70}$$

where v_e is the equilibrium velocity, *B* is the relaxation time, and c_0 is the sound speed of traffic. The material derivative appearing as two first terms on the left-hand side, demonstrates the Lagrangian motion of vehicles.

4. Aw and Rascle (2000) [119] presented a second-order anisotropic model. The isotropic behavior of all previous models leads to the unphysical backward motion of cars when they reach a dense cluster at the point of congestion. This case results in the creation of vertical queues.

$$\frac{\partial}{\partial t}(v + p(\rho)) + v\frac{\partial}{\partial x}(v + p(\rho)) = 0 \tag{6.71}$$

The p-term is a parameter similar to pressure and is proportional to ρ^γ where γ is a positive constant

6.5 POLAR FLUIDS

Polar fluids contain substructures that make their stress tensor asymmetrical. If the fluid is non-polar without any substructures, the only source of the moment is the torque of external body forces called the external angular momentum (Section 4.2). However, the intrinsic (internal) moments and the body torques can both lead to asymmetry of the stress tensor in polar fluids. Cowin (1974) [128] discussed that the fluids containing sub-substructures can be classified into the following items.

- **Polar fluid** contains non-deforming substructures. Consider that the definition of polar fluids should not be confused with the liquids such as water that have a polar molecular structure. In this context, water should be classified as a non-polar fluid. Such polarizations at the molecular level should be considered in microscale simulations such as molecular dynamics.

  The next point is that the polarity of a fluid has nothing to do with the rotationality of the flow field. Polar fluids has a non-zero dual vector of the stress tensor, $T_x = \varepsilon_{ikj}\tau_{jk}$. However, in rotational flows, the dual vector of the velocity gradient tensor (proportional to vorticity) is non-zero.

- **Dipolar fluids** is a category of fluids with deformable substructures, which their motion is directed by the velocity gradient tensor of the fluid flow.

- **Micropolar fluid** is a general form of polar fluids, including micro-structures such as rigid or sometimes deformable randomly oriented particles with different shapes suspended in a viscous fluid. Applications are nanofluids, granular flows containing large particles, and particulate flows. The difference between polar and micropolar fluids is the absolute rigidity of particles in polar fluids.
- **Ferromagnetic or ferroelectric fluid:** Ferrofluids or magnetic colloids are a kind of polarfluids that are sensitive to the electric or magnetic field. There is a base fluid containing nanometer-sized magnetic/electric particles as a suspension.
- **Magneto/electro-rheological fluids:** The structure of magneto-rheological fluids (MRF) is similar to that of ferrofluid. The main difference is the size of magnetic/electric particles in the fluid. The size of particles in the MRF fluids is at least one order of magnitude greater than the particles in ferrofluids.
- **Smart fluids** is a kind of magneto-rheological fluids with tunable parameters such as viscosity or surface tension coefficient when a magnetic or electric field is applied.

Conservation laws: Start from the conservations of mass, linear momentum, and angular momentum,

$$
\frac{D\rho}{Dt} + \rho\nabla\cdot\mathbf{u} = 0
$$

$$
\rho\frac{D\mathbf{u}}{Dt} = -\nabla p + (\lambda+\mu)\nabla(\nabla\cdot\mathbf{u}) + \mu\nabla^2\mathbf{u} + \mathbf{F}
$$

$$
\rho\frac{D}{Dt}(\mathbf{l}+\mathbf{r}\times\mathbf{u}) = \rho\mathbf{q}+\mathbf{r}\times\mathbf{F}+\mathbf{r}\times\nabla\cdot\boldsymbol{\tau}+\mathbf{T}_x+\nabla\cdot\mathbf{C} \tag{6.72}
$$

where $\mathbf{l}$ is the angular momentum per unit mass, $\mathbf{C}$ is the moment or couple stress tensor, $\mathbf{q}$ is the body torque, $\mathbf{T}_x$ is the dual vector of the stress tensor, and the left-hand side is the rate of change of total angular momentum which is the sum of internal and external angular momentums. $\mathbf{C}$ is a tensor similar to the stress tensor in the linear momentum law.

Computing the cross product of $\mathbf{r}$ and the Navier–Stokes equation[5] and subtracting the results form the mentioned angular momentum conservation law, we obtain the following equation presented by Łukaszewicz (1999) [57] for polar fluids,

$$
\rho\frac{D\mathbf{l}}{Dt} = \rho\mathbf{g}+\nabla\cdot C+T_x \rightarrow \rho I\frac{D\boldsymbol{\omega}}{Dt} = \rho\mathbf{q}+\nabla\cdot\mathbf{C}+\mathbf{T}_x \tag{6.73}
$$

where $\boldsymbol{\omega}$ is the micro-rotation vector representing the angular velocity of rotation of a particle, I is the micro-inertia coefficient which generally is a second-rank tensor in anisotropic fluids, and C is the moment or the couple stress tensor. If $\mathbf{C} = \mathbf{q} = \boldsymbol{\omega} = 0$, Equation (6.73) reduces to $\mathbf{T}_x = 0$ that is a condition referring to the symmetry of the stress tensor. $\rho\mathbf{l}$ and $\rho\mathbf{r}\times\mathbf{u}$ are called the internal (intrinsic) and the external angular momentums, respectively.

Constitutive relations should be presented both for the stress tensor and the couple stress tensor,

$$\tau_{ij}^c = -p\delta_{ij} + \mu\left(\frac{\partial u_i}{\partial x_j} + \frac{\partial u_j}{\partial x_i}\right) + \lambda\frac{\partial u_k}{\partial x_k}\delta_{ij} + \mu_r\left(\frac{\partial u_j}{\partial x_i} - \frac{\partial u_i}{\partial x_j}\right) - 2\mu_r\varepsilon_{mij}\omega_m$$

$$C_{ij} = c_0\frac{\partial\omega_k}{\partial x_k}\delta_{ij} + c_d\left(\frac{\partial\omega_i}{\partial x_j} + \frac{\partial\omega_j}{\partial x_i}\right) + c_a\left(\frac{\partial\omega_j}{\partial x_i} - \frac{\partial\omega_i}{\partial x_j}\right)$$

where μ_r is the dynamic micro-rotation viscosity, c_0, c_a, c_d are constants of angular viscosity. Substituting these relations into the conservation of mass, linear momentum, angular momentum, and energy we have

$$\frac{D\rho}{Dt} + \rho\nabla\cdot\mathbf{u} = 0$$

$$\rho\frac{D\mathbf{u}}{Dt} = -\nabla p + (\lambda + \mu - \mu_r)\nabla(\nabla\cdot\mathbf{u}) + (\mu + \mu_r)\nabla^2\mathbf{u} + \mathbf{F} + 2\mu_r\nabla\times\mathbf{u}$$

$$\rho\mathbf{I}\frac{D\boldsymbol{\omega}}{Dt} = 2\mu_r(\nabla\times\mathbf{u} - 2\boldsymbol{\omega}) + (c_0 + c_d - c_a)\nabla(\nabla\cdot\mathbf{u}) + (c_a + c_d)\nabla^2\boldsymbol{\omega} + \rho\mathbf{g}$$

$$\rho\frac{\partial e}{\partial t} = \rho\Phi - \nabla\cdot\mathbf{q} + p\nabla\cdot\mathbf{u} \tag{6.74}$$

where Φ is the dissipation function revisited for polar fluids,

$$\rho\Phi = \lambda(\nabla\cdot\mathbf{u})^2 + 2\mu\boldsymbol{\varepsilon}:\boldsymbol{\varepsilon} + 4\mu_r(0.5\nabla\times\mathbf{u} - \boldsymbol{\omega})^2$$
$$+ c_0(\nabla\cdot\boldsymbol{\omega})^2 + (c_a + c_d)\nabla\boldsymbol{\omega}:\nabla\boldsymbol{\omega} + (c_d - c_a)\nabla\boldsymbol{\omega}:\nabla\boldsymbol{\omega}^T \tag{6.75}$$

To solve the set of governing equations, we need to determine the value of $\boldsymbol{\omega}$ on boundaries. The only new constant in the Navier–Stokes equation is μ_r, which represents the extent of deviation from classical equations.

The second law considerations: To check the validity of the constitutive relations, we have to write the second law for polar fluids. In this case, the following conditions are obtained (Łukaszewicz, 1999) [57]. The two first items correspond to the Stokes hypothesis for Newtonian fluids.

$$3\lambda + 2\mu \geq 0$$
$$\mu \geq 0$$
$$\mu_r \geq 0$$
$$c_d \geq 0$$
$$c_d + c_a \geq 0$$
$$3c_0 + 2c_d \geq 0$$
$$c_a + c_d \geq c_d - c_a \geq -(c_a + c_d) \tag{6.76}$$

6.6 TURBULENT FLOW MODELING

It is not easy to present a precise definition of turbulent flow. Maybe, it can be described as a randomly three-dimensional, transient motion of a fluid. Turbulence undoubtedly is the most challenging and mysterious issue in analyzing the motion of a fluid.[6] The presented form of the Navier–Stokes equation is theoretically valid for both laminar and turbulent flows. However, the primary form is not applicable for the simulation of high-Reynolds number turbulent flows. Many attempts have been made to present mathematical models to understand complex underlying mechanisms of turbulent flows originating from their chaotic nature (Q6.5).

To model the complicated behavior of turbulent flows, different approaches exist.

- The direct numerical simulation, based on the direct solution of the Navier–Stokes equations without any modeling or approximations. The DNS simulations have greatly helped us understand the behavior of turbulent flows.
- The large-eddy simulation, based on spatial decomposition (filtering) of small scales and modeling their effect using the sub-grid-scale models.
- The Reynolds averaging, based on the temporal decomposition of quantities dealing with the time-averaged form of the governing equations. This approach yields the Reynolds-averaged Navier–Stokes (RANS) equations.

Other commonly used modeling techniques for turbulent flows are the unsteady RANS equation (URANS), the partially-averaged Navier–Stokes equation (PANS), the detached-eddy simulation (DES), and the very large-eddy simulation (VLES).

6.6.1 TURBULENT MEAN FLOW (RANS)

Turbulent flow can be characterized as a flow containing a small-to-large spectrum of eddies. To eliminate the unsteadiness of random three-dimensional transient fluctuations originating from eddies, we intend to decompose the velocity field into the mean velocity part and the turbulent fluctuation part. The process of temporal averaging is called the Reynolds decomposition as follows,

$$
\begin{aligned}
\rho(t) &= \bar{\rho} + \rho'(t) \\
u(t) &= \bar{u} + u'(t) \\
v(t) &= \bar{v} + v'(t) \\
w(t) &= \bar{w} + w'(t) \\
p(t) &= \bar{p} + p'(t) \\
e(t) &= \bar{e} + e'(t) \\
c(t) &= \bar{c} + c'(t)
\end{aligned}
\tag{6.77}
$$

where the time-averaging is performed during the time interval T. The necessary condition for this kind of manipulation of the mean value of parameters is the ergodicity of the flow field. For a quantity like η,

$$
\bar{\eta} = \frac{1}{T} \int_{t_0}^{t_0+T} \eta(t)dt
\tag{6.78}
$$

Pay attention that the averaging period should be larger than the time-scale of turbulent fluctuations. Since the average of fluctuations is zero

$$
\overline{u'} = \frac{1}{T} \int_{t_0}^{t_0+T} (u - \bar{u})dt = \frac{1}{T} \int_{t_0}^{t_0+T} udt - \bar{u} = \bar{u} - \bar{u} = 0
\tag{6.79}
$$

we need to find a parameter to demonstrate the strength of turbulence. Consequently, the root-mean-square of velocity fluctuations has been proposed,

$$u_{rms} = \sqrt{\overline{u_i' u_i'}} \tag{6.80}$$

The ratio of $\frac{u_{rms}}{\bar{u}}$ is called the turbulence intensity and $K = \frac{1}{2}\overline{u_i' u_i'}$ is the turbulent kinetic energy. Some identities based on the Reynolds decomposition are

$$
\begin{aligned}
f &= \bar{f} + f' \to \overline{\bar{f}} = \overline{\bar{f} + f'} = \bar{f} + \overline{f'} \to \overline{f'} = 0 \\
\frac{\overline{\partial f}}{\partial x} &= \frac{\partial \bar{f}}{\partial x} \\
\overline{\overline{f}} &= \frac{1}{T}\int_t^{t+T} \bar{f}\, dt = \bar{f}\frac{1}{T}\int_t^{t+T} dt = \bar{f} \\
\overline{\bar{f}g} &= \frac{1}{T}\int_t^{t+T} \bar{f}g\, dt = \bar{g}\frac{1}{T}\int_t^{t+T} f\, dt = \overline{f}\bar{g} \\
\overline{f'\bar{g}} &= \frac{1}{T}\int_t^{t+T} f'\bar{g}\, dt = \bar{g}\frac{1}{T}\int_t^{t+T} f'\, dt = \bar{g}\times\overline{f'} = \bar{g}\times 0 = 0 \\
\overline{fg} &= \overline{(\bar{f}+f')(\bar{g}+g')} = \overline{\bar{f}\bar{g} + \bar{f}g' + f'\bar{g} + f'g'} = \overline{f}\bar{g} + 0 + 0 + \overline{f'g'} = \overline{f}\bar{g} + \overline{f'g'}
\end{aligned}
\tag{6.81}
$$

The incompressible form: Taking the average of incompressible form of the continuity, the momentum, and the energy equations, the incompressible form of the mean equations can be obtained as follows,

$$\frac{\partial(u-u')}{\partial x} + \frac{\partial(v-v')}{\partial y} + \frac{\partial(w-w')}{\partial z} = 0 \to \frac{\partial u'}{\partial x} + \frac{\partial v'}{\partial y} + \frac{\partial w'}{\partial z} = 0$$

$$\rho\frac{D\bar{u}}{Dt} = \rho\mathbf{g} - \nabla\bar{p} + \nabla\cdot\boldsymbol{\tau}^r, \quad \tau_{ij}^r = \mu\left(\frac{\partial \bar{u}_i}{\partial x_j} + \frac{\partial \bar{u}_j}{\partial x_i}\right) - \rho\overline{u_i'u_j'} \tag{6.82}$$

The energy and concentration equations read

$$\rho c_p\frac{D\bar{T}}{Dt} = -\frac{\partial q_i}{\partial x_i} + \overline{\Phi}$$

$$q_i = -k\frac{\partial \bar{T}}{\partial x_i} + \rho c_p\overline{u_i'T'}$$

$$\overline{\Phi} = \frac{\mu}{2}\left(\frac{\partial \bar{u}_i}{\partial x_j} + \frac{\partial u_i'}{\partial x_j} + \frac{\partial \bar{u}_j}{\partial x_i} + \frac{\partial u_j'}{\partial x_i}\right)^2$$

$$\frac{D\bar{c}}{Dt} = \frac{\partial}{\partial x_i}\left(D\frac{\partial \bar{c}}{\partial x_i} - \overline{u_i'c'}\right) \tag{6.83}$$

The last equation represents the averaged version of the concentration equation in which the last term $\overline{u_i'c'}$ is the mass transfer due to advection of turbulence and can be approximated by $-D_t \frac{\partial \bar{c}}{\partial x_i}$. D_t is turbulent coefficient of diffusion. A similar process should be performed to compute other new terms like $\overline{u_i u_j}$ called the turbulent stress tensor, $\overline{u_i' T'}$ the turbulent flux vector, and the averaged viscous dissipation $\overline{\Phi}$. Also, a transport equation can be obtained by averaging the mechanical energy equation as a governing equation for the turbulent kinetic energy.

The compressible form: Let us take the average of the compressible form of the Navier–Stokes equation. The average of the sum of terms is equivalent to the sum of their averages. So, the average operator breaks and goes over each derivative,

$$\frac{\partial \bar{\rho}}{\partial t} + \frac{\partial \overline{\rho u_i}}{\partial x_i} = 0$$
$$\frac{\partial \overline{\rho u_i}}{\partial t} + \frac{\partial \overline{\rho u_i u_j}}{\partial x_j} = -\frac{\partial \bar{p}}{\partial x_i} + \frac{\partial \overline{\tau_{ij}}}{\partial x_j} \qquad (6.84)$$

in which the averaging and the spatial/temporal derivatives have been interchanged. In this relation, we see the average of the product of three quantities because the density is a variable. If we expand the average of the product of three parameters, we face a high degree of complexity in a number of terms in the equation. To prevent such sophistication stemming from variation of density, a new decomposition called the Favre averaging has been introduced

$$\eta = \tilde{\eta} + \eta'' \qquad (6.85)$$

in which $\tilde{\eta} = \frac{\overline{\rho \eta}}{\bar{\rho}}$. The identities for the Favre averaging are

$$\tilde{f} = \frac{\overline{\rho f}}{\bar{\rho}} = \frac{\overline{\bar{\rho}\bar{f}} + \overline{\rho' f'}}{\bar{\rho}} = \bar{f} + \frac{\overline{\rho' f'}}{\bar{\rho}}$$

$$f'' = f - \tilde{f} = f - \bar{f} - \frac{\overline{\rho' f'}}{\bar{\rho}}$$

$$\overline{\rho f''} = 0$$

$$\overline{\tilde{f}\tilde{g}} = \overline{\tilde{f}\frac{\overline{\rho g}}{\bar{\rho}}} = \bar{\tilde{f}}\frac{\overline{\rho g}}{\bar{\rho}} = \tilde{f}\tilde{g}$$

$$\overline{\rho f g} = \overline{\rho(\tilde{f} + f'')(\tilde{g} + g'')} = \overline{\rho \tilde{f}\tilde{g}} + \overline{\rho g'' \tilde{f}} + \overline{\rho f'' \tilde{g}} + \overline{\rho f'' g''}$$
$$= \bar{\rho}\tilde{f}\tilde{g} + \overline{\rho g''}\tilde{f} + \overline{\rho f''}\tilde{g} + \overline{\rho f'' g''} = \bar{\rho}\tilde{f}\tilde{g} + 0 + 0 + \overline{\rho f'' g''} = \bar{\rho}\tilde{f}\tilde{g} + \overline{\rho f'' g''}$$

$$\overline{\rho f'' g''} = \overline{\bar{\rho}\widetilde{f'' g''}}$$

$$\overline{\rho f} = \overline{\rho f \times 1} = \bar{\rho}\tilde{f}\tilde{1} + \overline{\rho f'' 1''} = \bar{\rho}\tilde{f} \qquad (6.86)$$

Using the last identity, it is so easy to show that the averaged continuity equation is

$$\frac{\partial \bar{\rho}}{\partial t} + \frac{\partial \bar{\rho}\tilde{u}_i}{\partial x_i} = 0 \qquad (6.87)$$

and the averaged momentum equation is

$$\frac{\partial \overline{\rho(\tilde{u}_i + u_i'')}}{\partial t} + \frac{\partial \overline{\rho(\tilde{u}_i + u_i'')(\tilde{u}_j + u_j'')}}{\partial x_j} = -\frac{\partial \overline{(\bar{p} + p')}}{\partial x_i} + \frac{\partial \overline{(\overline{\tau}_{ij} + \tau_{ij})}}{\partial x_j} \rightarrow$$

$$\frac{\partial \bar{\rho}\tilde{u}_i}{\partial t} + \frac{\partial (\bar{\rho}\tilde{u}_i\tilde{u}_j + \overline{\rho u_i'' u_j''})}{\partial x_j} = -\frac{\partial \bar{p}}{\partial x_i} + \frac{\partial \overline{\tau}_{ij}}{\partial x_j} \tag{6.88}$$

and the equation of state is

$$\bar{p} = \bar{\rho}R\tilde{T} \tag{6.89}$$

6.6.2 FILTERED EQUATIONS (LES)

The filtering process: The filtered parameter $(\bar{f})$ and the filtering operator denoted by $\star$ are defined by the Leonard's convolution integral

$$\bar{f}(\mathbf{r},t) = \int_\Omega G(\mathbf{r}-\mathbf{r}',t-t')f(\mathbf{r}',t')dt'd\mathbf{r}' = G\star f \tag{6.90}$$

The filter is proposed to eliminate the turbulent structures smaller than a cut-off length Δ in spatial filtering, and shorter than a cut-off time τ in temporal filtering. The convolution integral adds the non-local behavior to the mathematical behavior of results. The cut-off parameter is hidden in the mathematical structure of the convolution kernel function G, and may be a function of the mesh size and the time-step size of the numerical solution. In the frequency domain, the filter omits the high-frequency scales. It should be mentioned that small eddies correspond to high wave numbers and vice versa. A similar filtering process using a convolution integral may be performed in particle-laden flows in which the effect of small particles is filtered [127].

Based on the definition of filters, the filtered equations in the large-eddy simulation framework can be constructed by applying the following decomposition to all parameters,

$$f'(\mathbf{r},t) = f(\mathbf{r},t) - \bar{f}(\mathbf{r},t) = (1-G)\star f \tag{6.91}$$

The primmed term denotes the non-resolved part of f.

This way, we are omitting the energy of small structures in the energy spectrum. Then, we have to compensate for the effect of the removed scales by adding the sub-grid-scale terms to the filtered equation. If the kernel function G does not depend on space coordinate $(\mathbf{r})$ and remains identical at different locations, it is called a homogenous filter. Then, the consistency condition can be derived for a homogenous filter when $f \equiv c$

$$1 = \int_\Omega G(\mathbf{r}-\mathbf{r}',t-t')dt'd\mathbf{r}' \tag{6.92}$$

The new decomposition (filtering) has the following properties

$$\begin{aligned}
\overline{\overline{f}} &\neq \overline{f} \rightarrow G \star G \star f \neq G \star f \\
\overline{f+g} &= \overline{f} + \overline{g} \\
\overline{\frac{\partial f}{\partial x}} &= \frac{\partial \overline{f}}{\partial x} \\
\overline{f'} &= G \star (1-G) \star f \neq 0 \\
\overline{\overline{f}g} &\neq \overline{f}\,\overline{g} \\
\overline{fg} &\neq \overline{f}\,\overline{g}
\end{aligned} \tag{6.93}$$

Regardless of averaging in RANS models, the first identity implies that when you filter a parameter twice, the output of the second filtering differs from the result of the first filtering. It should be noted that the third property which is about commuting the filter with the derivative is just valid for homogeneous filters.

Types of filters: One of the most common homogenous spatial filters is the box filter (sometimes called the top hat filter) as a local filter in the physical space,

$$G(\mathbf{r}-\mathbf{r}') = \frac{1}{\Delta} \quad if : |\mathbf{r}-\mathbf{r}'| \leq \frac{\Delta}{2} \tag{6.94}$$

otherwise, G is zero. Δ is the width of the filter. In the one-dimensional case,

$$\overline{f}(x) = \frac{1}{\Delta} \int_{x-\frac{\Delta}{2}}^{x+\frac{\Delta}{2}} f(x')dx' \tag{6.95}$$

Equation (6.95) demonstrates that this filter is equivalent to the spatial averaging of the function in a small interval around the point x. Using the top hat filter, the high-frequency fluctuations around the average of f in an interval are all omitted and the function becomes smooth. Again, it is clear that $\overline{\overline{f}} \neq \overline{f}$, since the smoothing procedure never stops, and each smoothed function differs from the next smoothed functions at the next steps.

Other famous filters are the Gaussian filter, the sharp spectral/cutoff filter, the integral implicit filter, and implicit numerical filtering. The implicit filtering is popular in numerical methods due to its simplicity. In such filters, we replace the filtering step with finite-difference discretization with the filter width being equal to twice the grid spacing (h),

$$\begin{aligned}
\frac{d\overline{u}}{dx} = \frac{d}{dx}\left[\frac{1}{2h}\int_{x-h}^{x+h} u(x')dx'\right] &= \frac{d}{dx}\left[\frac{F(x+h)-F(x-h)}{2h}\right] \\
&= \frac{u(x+h)-u(x-h)}{2h}
\end{aligned} \tag{6.96}$$

where F is the primitive function of u. The implicit filtering is inhomogeneous for non-uniform meshes.

The incompressible filtered equation: Applying the above-mentioned filter to the incompressible form of the Navier–Stokes and energy equations

$$\frac{\partial \bar{u}_i}{\partial x_i} = 0$$

$$\frac{\partial \bar{u}_i}{\partial t} + \frac{\partial \overline{u_i u_j}}{\partial x_j} = -\frac{1}{\rho}\frac{\partial \bar{p}}{\partial x_i} + v\frac{\partial^2 \bar{u}_i}{\partial x_j \partial x_j}$$

$$\frac{\partial \bar{T}}{\partial t} + \frac{\partial \overline{u_i T}}{\partial x_j} = \frac{\partial}{\partial x_j}\alpha\frac{\partial \bar{T}}{\partial x_j} \tag{6.97}$$

The mathematical structure of the filtered continuity equation is similar to the averaged continuity equation in the RANS models. Since $\overline{fg} \neq \bar{f}\bar{g}$, we have to add a new term ($\frac{\partial \bar{u}_i \bar{u}_j}{\partial x_j}$ and $\frac{\partial \bar{u}_i \bar{T}}{\partial x_j}$) to the both sides to obtain

$$\frac{\partial \bar{u}_i}{\partial t} + \frac{\partial \bar{u}_i \bar{u}_j}{\partial x_j} = -\frac{1}{\rho}\frac{\partial \bar{p}}{\partial x_i} + v\frac{\partial^2 \bar{u}_i}{\partial x_j \partial x_j} + \frac{\partial \bar{u}_i \bar{u}_j}{\partial x_j} - \frac{\partial \overline{u_i u_j}}{\partial x_j}$$

$$\frac{\partial \bar{T}}{\partial t} + \frac{\partial \bar{u}_i \bar{T}}{\partial x_j} = \frac{\partial}{\partial x_j}\alpha\frac{\partial \bar{T}}{\partial x_j} + \frac{\partial \bar{u}_i \bar{T}}{\partial x_j} - \frac{\partial \overline{u_i T}}{\partial x_j} \tag{6.98}$$

So,

$$\frac{\partial \bar{u}_i}{\partial t} + \frac{\partial \bar{u}_i \bar{u}_j}{\partial x_j} = -\frac{1}{\rho}\frac{\partial \bar{p}}{\partial x_i} + v\frac{\partial^2 \bar{u}_i}{\partial x_j \partial x_j} - \frac{\partial \tau^s_{ij}}{\partial x_j}$$

$$\frac{\partial \bar{T}}{\partial t} + \frac{\partial \bar{u}_i \bar{T}}{\partial x_j} = \frac{\partial}{\partial x_j}\alpha\frac{\partial \bar{T}}{\partial x_j} + \frac{\partial q^s_j}{\partial x_j} \tag{6.99}$$

where

$$\tau^s_{ij} = \overline{u_i u_j} - \bar{u}_i \bar{u}_j$$

$$q^s_j = \overline{u_i T} - \bar{u}_i \bar{T} \tag{6.100}$$

τ^s_{ij} and q^s_j are the subgrid-scale tensors or the residual-stress tensors, which should be modeled. These tensors encounter the effect of the missing filtered structures. The role of such tensors is similar to that of the Reynolds stress tensor in the RANS models. The modified filtered pressure and the modified residual stress tensor are defined by adding and subtracting $\frac{1}{3}\tau^s_{kk}\delta_{ij}$ from the equation to make the subgrid-scale tensor traceless similar to the filtered strain-rate tensor.

$$\tilde{p} \equiv \bar{p} + \frac{1}{3}\tau^s_{kk}\delta_{ij}$$

$$\tau^m_{ij} \equiv \tau^s_{ij} - \frac{1}{3}\tau^s_{kk}\delta_{ij} \tag{6.101}$$

So,

$$\frac{\partial \bar{u}_i}{\partial t} + \frac{\partial \bar{u}_i \bar{u}_j}{\partial x_j} = -\frac{1}{\rho}\frac{\partial \tilde{p}}{\partial x_i} + v\frac{\partial^2 \bar{u}_i}{\partial x_j \partial x_j} - \frac{\partial \tau^m_{ij}}{\partial x_j} \tag{6.102}$$

As a good exercise, try to derive the filtered form of the nonlinear viscous Burgers' equation.

The compressible filtered equation: Similar to the procedure in the RANS equations, the Favre decomposition is used and the filtered continuity and momentum equations are

$$\frac{\partial \overline{\rho}}{\partial t} + \frac{\partial \overline{\rho}\tilde{u}_i}{\partial x_i} = 0$$

$$\frac{\partial \overline{\rho}\tilde{u}_i}{\partial t} + \frac{\partial \overline{\rho}\tilde{u}_i\tilde{u}_j}{\partial x_j} = -\frac{\partial \overline{p}}{\partial x_i} - \frac{\overline{\rho}\partial \tau_{ij}^m}{\partial x_j} + \frac{\partial \overline{\sigma}_{ij}}{\partial x_j}$$

where

$$\sigma_{ij} = 2\mu \tilde{S}_{ij} - \frac{2}{3}\mu \delta_{ij} \tilde{S}_{kk}$$

$$\tau_{ij}^s = \overline{\rho}(\widetilde{u_iu_j} - \tilde{u}_i\tilde{u}_j)$$

$$\tau_{ij}^m = \tau_{ij}^s - \frac{1}{3}\tau_{kk}\delta_{ij} \tag{6.103}$$

The subgrid-scale stress tensor: Estimating the SGS tensor is the heart of any LES approach. An accurate SGS model should statistically predict the correct energy spectrum especially for energy carrying structures, the interaction of large and small-scale eddies, the dissipation of kinetic energy, and also should be Galilean invariant.

To evaluate the accuracy of an SGS model, we can perform the priori or the posteriori tests. In the posteriori test, we already have precise results obtained from the DNS study or experiments. The exact results should be filtered to compute precise values of the SGS tensor based on its original definition without employing any model. Then, we recalculate the SGS tensor, this time, using the exact filtered velocities and our own SGS model. Finally, two results can be compared. In the posteriori test as a top–down approach, you do not need to perform an LES simulation. In the priori test (down–top approach), the results of an existing LES simulation are compared with those obtained from the DNS or experiments.

The Germano and Leonard decompositions are two common SGS tensor decompositions that may help us to propose an accurate and flexible SGS model. In these decompositions, the SGS tensor has been broken into the Leonard part for large scales, the cross term for large–small interactions, and the Reynolds-like sub-filter stresses. Despite Leonard's decomposition, Germano's decomposition is Galilean invariant.

Generally, we can classify different SGS models into four categories.

1. **Static/dynamic functional models:** The functional models focus on the dissipation of energy based on the Boussinessq hypothesis. The deviatoric part of the subgrid-scale tensor is modeled based on the eddy-viscosity concept,

$$\tau_{ij}^s - \frac{1}{3}\delta_{ij}\tau_{kk} = 2\nu_t \overline{S}_{ij}$$

$$\overline{S}_{ij} = \frac{1}{2}\left(\frac{\partial \overline{u}_i}{\partial x_j} + \frac{\partial \overline{u}_j}{\partial x_i}\right) \tag{6.104}$$

This way, a new turbulent viscosity v_t is added to the fluid's molecular viscosity. The celebrated Smagorinsky–Lilly model was presented based on Prandtl's mixing length,

$$v_t = C_s \Delta^2 \sqrt{2\overline{S}_{ij}\overline{S}_{ij}} \tag{6.105}$$

where Δ equals $Max(\Delta_1, \Delta_2, \Delta_3)$ or $(\Delta_1\Delta_2\Delta_3)^{\frac{1}{3}}$ or $(\frac{\Delta_1^2 + \Delta_2^2 + \Delta_3^2}{3})^{\frac{1}{2}}$ and represents the cut-off filter size. C_s is a positive constant that prevents the accumulation of kinetic energy in small scales and the instability of the solution procedure. Regardless of the RANS models, Equation (6.105) proves that if the cut-off length goes to zero, the turbulent viscosity vanishes.

In static models, the parameter C_s is a constant not a function of the Reynolds number, the Richardson number, y^+, or $\frac{L}{\Delta}$. The magnitude of C_s widely varies from 0.06 for shear layers to 0.25 for homogeneous isotropic turbulence. To deal with this shortage, we may use artificial damping functions such as the Van Driest equation or the dynamics functional models.

Besides the robustness of the Smagorinsky model, the backscattering from small to large scales is absent in such models, and C_s is non-zero in laminar parts of the domain. In dynamic models such as Germano's model, the constant of the model is computed as a function of the spatial coordinate and time and becomes zero on walls. The basic idea behind the dynamic models is to calibrate the model by double-filtering using a wider filter. With this strategy, we can compute the model constants without the need for any external information. We call these models dynamic, since the turbulent viscosity varies depending on turbulent flow dynamics, and may generate negative values of v_t. To limit the range of variation of v_t, some smoothing ensemble averaging over homogeneous directions of neighboring cells is needed for computed values on different points. Despite the static model, the dynamic turbulent viscosity is zero in laminar flow regions.

2. **Bardina self-similar structural model:** Structural models have been presented based on the concept of estimating the unfiltered scales from filtered scales referring to the self-similarity of turbulent flows (scale similarity hypothesis). The main idea behind the structural models is an extrapolation of unresolved scales near the cut-off frequency from smaller resolved scales. In self-similar models, we do not need an external predefined framework for the model.

3. **Mixed models:** The structural models are not dissipative and may become unstable during the numerical simulations. In the Bardina model, the SGS tensor is just equivalent to the Leonard tensor in Germano decomposition. To make the structural models dissipative, the Smagorinsky model is used to compute the cross tensor and the Reynolds-like tensor in the decomposition. So, the final model is a combination of the Bardina and the Smagorinsky models.

4. **Other models:** In the detached eddy simulation (DES) models, the turbulent viscosity is computed based on the RANS models in near-wall regions, and the large-eddy simulation SGS models in free-shear layers. This concept acts as a multi-mode RANS-LES scheme with a wall function.

Some of the other models are the dynamic localization model (based on an integral equation), the Lagrangian averaging (averaging along pathlines with memory effects), the structure-function models, the approximate deconvolution method, the dynamic divergence model (directly estimating the divergence of the subgrid-scale tensor), the fourth-order viscosity model (containing a term proportional to C_s^4 called the hyper-viscosity), the multiscale models, the RNG-LES models (based on the renormalized group theory), and the dynamic global coefficient model.

6.7 BIOLOGICAL HEAT TRANSFER

The structure of living biological tissues is so complex. There are numerous capillary and larger vessels with three-dimensional curved geometries. Another complication is the blood flow in vessels and heat generation due to metabolism. Here, we are going to introduce continuum models for thermal biological applications to model all these complex phenomena in a single framework. Predicting temperature distribution inside biological media is important in applications such as laser therapy, cryotherapy, thermo-regulation, inflammation, ablation, and hyperthermia treatment (Q6.6).

Figure 6.2 illustrates a schematic of the vascular network near the peripheral skin layer. To present an ab initio simulation without the need for a continuum model, we have to take into account the anatomy of the vascular network, the blood perfusion in the capillary bed, mass transfer between the vessel and the tissue, the temperature difference between the blood and the surrounding tissue, the local distribution of the tissue temperature, the deformable geometry of vessels, the convection of heat due to motion of blood in the vessel, the blood-tissue thermal interaction, and metabolism.

The capillary network consists of vessels with a diameter less than 60 microns. Metabolism is the chemical reaction in living organs that leads to the production of energy. The perfusion is the phenomenon resulting from the passage of hot blood

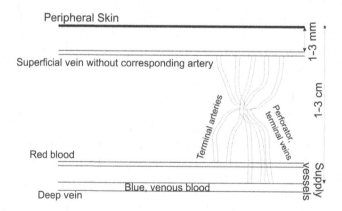

Figure 6.2   Schematic of blood vessels in living tissues.

through the circulatory lymphatic system toward an organ to deliver blood to the capillary bed in tissue. The rate of perfusion is a function of the level of a human's normal daily activity, pathology, anatomy, environmental conditions, and physiological characteristics.

To compute the degree of burn injury (D) in tissues during the heating period $(0-t)$, an integral transformation has been proposed

$$D = \int_0^t A e^{-\frac{E}{RT(t')}} dt'$$

where $A = 3.1 \times 10^{98} s^{-1}$ is the scaling factor and $E = 6.27 \times 10^8 J/kmol$ is the activation energy [129].

The Pennes' model: Pennes (1948) [130] performed experiments on human forearm subjects, and presented his well-known equation for thermal biological cases.

$$\rho c \frac{\partial T}{\partial t} = \nabla(k\nabla T) + (\rho c)_b \omega_b (T_a - T) + q_{met} + q_{source/laser} \qquad (6.106)$$

where the subscript "b" refers to blood, q_{met} is the heat generated per unit volume due to metabolism, $q_{source/laser}$ is the source term like the laser irradiation, T_a is the temperature of the arterial blood, T is the temperature of the surrounding tissue, and ω is the perfusion rate [131].

The main structure of the Pennes-based heat equation is similar to the classical heat energy, but two new terms have been added. The first one is linear in temperature and adds the perfusion effects. The other one is the metabolism heat generation, which may be supposed to be homogenous. It is expected that the energy equation derived based on Fourier's law is no longer valid for such sophisticated media.

In Pennes' equation, simultaneous effects of radiation as the source term, convection as the perfusion term, and conduction exist. The main point about the Pennes' equation is that this model does not use geometrical information about the vessels such as the radius of vessels. Also, other details such as the counter-current of the flow in arteries and veins have been ignored. The mass transfer between the vessels and the tissue is not included as well.

The Pennes' paradigm tries to model the effect of all vessels as a single continuum. Maybe, some larger vessels have to be modeled separately. We need a criterion to distinguish the vessels that are participating in heat transfer. The vessels with a diameter in the range of 50 to 500 microns are thermally significant. In such vessels, the thermal equilibrium length is in the order of the length of the vessel. Here, the main assumptions of Pennes' model are listed.

1. Using Fourier's law to compute the conduction term.
2. Presented for human's forearm.
3. The perfusion rate is uniform.
4. The metabolism heat source is uniform.
5. The mass transfer is ignored.
6. Ignoring geometrical details.

The Chen–Holmes model modifies the Pennes' model to take into account other details such as the number of vessels and their radius.

$$
\begin{aligned}
\rho c \frac{\partial T}{\partial t} &= \nabla(k\nabla T) + (\rho c)_b \omega_b^*(T_a^* - T) - (\rho c)_b \omega_b^* \mathbf{u} \cdot \nabla T \\
&+ \nabla(\mathbf{k}_p \cdot \nabla T) + q_{met} + q_{source}
\end{aligned}
\tag{6.107}
$$

where $\mathbf{k}_p = f(n, (\rho c)_b, R_b, \mathbf{u}, \gamma, L)$ is the perfusion conductivity tensor, n is the number of vessels, R_b is the vessel radius, $\mathbf{u}$ is the velocity of blood in the vessel, the subscript $*$ is related to the volume under consideration, γ is the angle between ∇T and the vessel [132].

The Weinbaum–Jiji–Lemons model (WJL): The WJL model is one of the vascular-based models, which can model the counter-current vasculature, the vein-artery pair, and the contribution of perfusion to heat transfer as a porous medium. Assumptions of the WJL model are neglecting the lymphatic loss, which leads to the presumption of the same mass flow for arteries and veins, and identical diameters for veins and arteries. The W–J model is constructed based on the three following equations.

$$
\begin{aligned}
(\rho c)_b \pi r_b^2 \mathbf{u} \frac{dT_a}{ds} &= -q_a \\
(\rho c)_b \pi r_b^2 \mathbf{u} \frac{dT_v}{ds} &= -q_v \\
\rho c \frac{\partial T}{\partial t} &= \nabla(k\nabla T) + ng(\rho c)_b (T_a - T_v) - n\pi r_b^2 (\rho c)_b \mathbf{u} \frac{d(T_s - T_v)}{ds} + q_{met}
\end{aligned}
\tag{6.108}
$$

where the two first equations deal with heat transfer to the thermally significant arteries and veins, respectively. The second and the third terms in the right-hand side of the third equation are the capillary bleed-off energy exchange as a term similar to the perfusion rate in Pennes' model, and the net heat exchange between the tissue and the pair. T_v is the temperature of the tissue surrounding the artery-vein pair [133].

The simplified Weinbaum–Jiji (W–J) model: The WJL model is a combination of three equations that are not easy to implement. The W–J model is the simplified form of the WJL model.

$$
\begin{aligned}
\rho c \frac{\partial T}{\partial t} &= \nabla(k_e \nabla T) + q_{met} \\
k_e &= k\left[1 + \frac{2[(\rho c)_b \pi r_b^2 \mathbf{u} \cos \gamma]}{\sigma k^2}\right] \\
q_a &= q_v = \sigma k(T_a - T_v)
\end{aligned}
\tag{6.109}
$$

where σ is the shape factor. The last equation implies that the heat is mostly conducted from the paired artery to the vein. In the W–J model, the imperfect counter-current heat exchange is included as an effective conductivity tensor [134].

The DPL-based bioheat model: Zhou et al. (2009) presented the DPL-based model of thermal biological energy equation [135],

$$\rho c \left(\frac{\partial T}{\partial t} + \tau_q \frac{\partial^2 T}{\partial t^2} \right) = \nabla(k\nabla T) + \tau_T \frac{\partial}{\partial t} \nabla(k\nabla T) + (\rho c)_b \omega_b (T_a - T)$$

$$+ \left(q_{met} + q_{source/laser} \right) + \tau_q \frac{\partial}{\partial t} \left(q_{met} + q_{source/laser} \right)$$

(6.110)

If $\tau_T = 0$, the model reduces to the CV-based thermal wave bioheat model. The name comes from the wave-like behavior of the CV-based energy equation due to its hyperbolic nature [136, 137].

6.8 MULTIPHASE FLOW MODELING

If a combination of two phases such as gases, liquids, solids, and plasma or even two distinct materials of the same phase[7] appear, the flow configuration is two-phase. Multiphase flows in some special applications like rheological fluids are called complex fluids.

6.8.1 CLASSIFICATIONS

Differences between gas and vapor: Vapor is not a new phase of matter, but an unstable permutation of the gaseous phase with properties between those of gases and liquids. Vapor can be created by heating a liquid or a solid. Both of them may be smelly and toxic. Vapor is visible but gases like air are fortunately invisible! Gases expand to fill the space, but the vapor is just locally distributed and fills a part of the space. So, concepts such as relative humidity can be defined just for vapors.

The running water vapor molecules from the free-surface of water form clusters, which still contain hydrogen bonds that makes the vapor visible. However, the intermolecular forces in gas particles are weak. In conclusion, gravity affects vapors. So, they can sediment. On contrary, gravity has a negligible effect on gaseous flows due to their small density. The supercritical gas appears beyond the critical pressure and temperature. In that situation, the vapor is converted to gas, the free-surface fades out, and the surface tension coefficient becomes zero.

The phase-change process includes melting (solid to liquid), solidification or freezing (liquid to solid), condensation (gas/vapor to liquid), sublimation (solid to gas), deposition (gas/vapor to solid), boiling, vaporization or cavitation (liquid to gas/vapor), ionization (gas to plasma), and recombination (plasma to gas).

Differences between boiling and vaporization: Vaporization occurs when a high-velocity stream of gas moves over free-surface of a liquid. During this process, the water particles near the free-surface run out of the liquid phase and enter the vapor phase. Vaporization is an invisible, silent, and slow surface process, while boiling is a visible, rapid phase-change of liquid to vapor with bulk acoustic effects.

Vaporization can occur at any temperature, but boiling takes place just at the boiling temperature of the liquid. Vaporization is a microscopic molecular phenomenon, while boiling is a macroscopic thermal process with the bulk motion of bubbly fluid formed at nucleation sites as well as the surface tension effects in the bubbles. However, the velocity field of a boiling liquid is a complex, multi-regime, highly non-uniform field containing fluid jets, hydrodynamic instabilities, gravity, and surface-tension effects.

Classification and regimes: Multiphase flows can be categorized based on the type of phases interacting with each other. The density ratio $\gamma = \frac{\rho_D}{\rho_B}$ can be defined for each of these flows, where ρ_D and ρ_B are densities of the dispersed and the base phases, respectively.

- Liquid–gas flows such as bubbly flows (gas-in-liquid two-phase flow), droplet dynamics (liquid-in-gas two-phase flow), free-surface or interfacial flows (two immiscible liquids or a gas over a liquid), boiling, condensation, sprays, mist flows, slug flows, combustion chambers, atomization, cavitation, evaporators, boilers, offshore structures, and jet cooling.
- Solid–liquid flows ($\gamma \simeq 1$) such as nanofluids, mineral liquids, slurry flows, erosion, cyclones, blood, sedimentation in rivers and dams, groundwater flows, fibrous flows, and hydro-transport with dispersed solid particles in a continuous liquid phase.
- Gas–solid flows ($\gamma \simeq 1000$) such as particle transport in atmospheric flows, aerosol dispersion, granular flows, particle-laden flows, environmental flows, pollution, and dynamics of suspended particles.
- Multiphase flows include the interaction of more than two phases such as the moving bed free-surface gas-solid-liquid flows, solid catalysts in liquid-gas reactions, solid particles in two-fluid flows, or the three-phase oil/gas/water flows in the petroleum industry.

Regimes of multiphase flows are: (1) the separated flows like stratified flows with a distinct interface between two phases and the slug flow including a large region of the gaseous phase in a liquid, (2) the dispersed flow with the disconnected distribution of particles in a base fluid and (3) the mixed flow which is a combination of the separated and the dispersed regimes.

Another key dimensionless number in two-phase flows is the Stokes number (St) defined as the ratio of the particle relaxation time or the stopping time and the flow timescale. If the Stokes number increases, the particle stopping time also increases and the motion of the particle is hardly affected by the flow structures such as eddies. Conversely, if the stopping time is a small quantity, the particle precisely follows the curvature of streamlines of the base flow. This situation is called the full advection case. Consequently, by increasing the Stokes number, the motion of particles becomes independent of the flow velocity field and just obeys the trajectory of the particle starting from an initial velocity.

The volume occupied by each phase can be described by the volume fraction parameter, α_i. Another classification based on the particle loading as the ratio of effective density of two phases, $L = \frac{\alpha_D \rho_D}{\alpha_B \rho_B}$ is (1) small-particle loading (dilute) multiphase flows that show one-way coupling. In this case, just the base flow controls the motion of the second phase. (2) intermediate loading flows, which have two-way coupling, and the drag force applied to the dispersed phase reduces the flow momentum and changes the turbulence level of the base fluid. In this regime, the Stokes number plays the key role. (3) High-particle loading (dense) flows again have two-way coupling, but there are other aspects such as variation of viscous stresses and pressure due to the motion of particles.

6.8.2 MODELING APPROACHES

All multiphase flow models can be used for low-particle loading cases. The selection of the model for intermediate-particle loading is based on the Stokes number.

1. **Fully Eulerian methods**
   - **The Eulerian–Eulerian model:** In this method, both the dispersed and the base phases are treated as a fluid by defining a volume fraction for each phase. The interaction between the two phases and between walls and particles should be considered. This method usually generates acceptable results for a high-volume fraction of the dispersed phase. A set of complete governing equations is solved for each phase, and the results of phases are coupled through inter-phase linking coefficients.

     In Eulerian–Eulerian models, the multiphase flow properties can be obtained by performing an ensemble or time-averaging process. For example, the velocity of phase i is $\mathbf{u}_i = \frac{1}{\alpha_i \forall} \int_\forall \mathbf{u}_i \beta_i d\forall$ in which β_i is zero outside phase i, and equals 1 inside it. The continuity equation should be written for each phase

     $$\frac{\partial}{\partial t}(\alpha_i \rho_i) + \nabla \cdot (\alpha_i \rho_i \mathbf{u}_i) = \sum_{j=1}^{N} m_{ji} \qquad (6.111)$$

     where m_{ij} is the rate of mass exchanged between phases i and j among N phases, and the net mass exchange is zero. The momentum equation has been modified by adding volume fractions and a source term $\mathbf{F}_D$

     $$\frac{\partial}{\partial t}(\alpha_i \rho_i \mathbf{u}_i) \quad + \quad \nabla \cdot (\alpha_i \rho_i \mathbf{u}_i \mathbf{u}_i) = -\alpha_i \nabla p_i + \nabla \cdot (\alpha_i \mu_i \mathbf{D}_i) + \alpha_i \rho_i \mathbf{g} + \mathbf{F}_{Di}$$

     $$+ \quad \sum_{j=1}^{N} (\mathbf{P}_{ji} + m_{ji} \mathbf{u}_i]) \qquad (6.112)$$

     $\mathbf{F}_{Di}$ is the interfacial force exerted to phase i and should be modeled. Consider that the indices i and j refer to phases and should and have nothing to do with the rules of index notation. The last term in the right-hand side shows the force exchange between two phases and the momentum transfer due to mass transport, and $\mathbf{P}_{ji} = K_{ji}[\mathbf{u}_j - \mathbf{u}_i] = -\mathbf{P}_{ij}$. In Eulerian–Eulerian multi-fluid models, the aforementioned set of equations should be solved for each phase. Due to the continuous nature of the discrete phase, such methods are well-appropriate for high-volume fraction multiphase flows. Similar equations may be written for the energy equation and the mass transport of species.

- **The single-equation mixture model** is a simplified version of the Eulerian–Eulerian model for low-Stokes numbers in which only one momentum equation is solved for the base fluid. Then, a concentration transport equation is added to model the motion of all other secondary phases. The interface of the separated streams can be captured by applying proper boundary conditions. The mixture model is appropriate for low and moderate particle loadings. The continuity equation for the mixture is

$$\frac{\partial \rho_m}{\partial t} + \nabla \cdot (\rho_m \mathbf{u}_m) = 0 \qquad (6.113)$$

The momentum mixture equation is

$$\frac{\partial \rho_m \mathbf{u}_m}{\partial t} + \nabla \cdot (\rho_m \mathbf{u}_m \mathbf{u}_m) = -\nabla p_m + \rho_m \mathbf{g} + \nabla \cdot (\mu_m \mathbf{D}_m) + \mathbf{F}_{Source}$$
$$+ \nabla \cdot \sum_{k=1}^{N} \alpha_k \rho_k \mathbf{u}_k^d \mathbf{u}_k^d \qquad (6.114)$$

The mixture density, viscosity, pressure gradient, and velocity are computed using the following relations

$$\rho_m = \sum_{i=1}^{N} \alpha_i \rho_i$$

$$\mu_m = \sum_{i=1}^{N} \alpha_i \mu_i$$

$$\nabla p_m = \sum_{i=1}^{N} \alpha_i \nabla p_i$$

$$\mathbf{u}_m = \frac{1}{\rho_m} \sum_{i=1}^{N} \alpha_i \rho_i \mathbf{u}_i$$

Based on the first equation, the density is a variable, and the mass fraction is $\frac{\alpha_i \rho_i}{\rho_m}$. The drift or the diffusion velocity is the relative velocity of phase i with respect to the velocity of the mixture $\mathbf{u}_i^d = \mathbf{u}_i - \mathbf{u}_m$. To close equations, the distribution of the volume fraction of each of N phases α_i should be computed using

$$\frac{\partial \alpha_i \rho_i}{\partial t} + \nabla \cdot (\rho_i \alpha_i \mathbf{u}_m) = -\nabla \cdot (\alpha_i \rho_i \mathbf{u}_i^d) \qquad (6.115)$$

- **The volume of fluid (VOF) method** is a simple Eulerian–Eulerian model for free-surface capturing in separated flows including surface tension effects. Similar to the mixture model, a fraction function is defined α_i, which is a continuous function varying between 0 for a fully gaseous condition and 1 for fully liquid case. The transport equation for the fraction function is

$$\frac{\partial \alpha_i}{\partial t} + \mathbf{u}\nabla \cdot \alpha_i = 0 \tag{6.116}$$

where $\sum_{i=1}^{N} \alpha_i = 1$ and $\rho = \sum_{i=1}^{N} \rho_i \alpha_i$. There are different categories of the VOF method like the donor–acceptor formulation and the line techniques [138].

- **The flame advancement** is similar to the VOF method in premixed reactions where the low-temperature reactants can propagate into the high-temperature products. So, to predict the evolution of the premixed flames in combustion problems, another transport equation should be solved to compute the flame motion called the G-equation,

$$\frac{\partial G}{\partial t} + \mathbf{u}\cdot\nabla G = U_F|\nabla G| \tag{6.117}$$

where U_F is the flame velocity, $G = 0$ determines the boundary of the flame, $G < 0$ denotes the unburned mixture, and $G > 0$ corresponds to the gaseous products [139].

- **The Buckley–Leverett equation in porous media** is a two-phase one-dimensional hyperbolic-type transport equation for modeling two-phase flow in porous media engaging with immiscible incompressible phases,

$$\frac{\partial F}{\partial t} + \frac{\partial}{\partial x}\left(\frac{Q}{\phi A}f(F)\right) = \phi\frac{\partial F}{\partial x} \tag{6.118}$$

where ϕ is the porosity, F denotes the saturation of the progressing phase, f demonstrates the relative mobilities of the two phases, Q is the flow rate, and A is the cross sectional area [141].

2. **The Eulerian-Lagrangian models or the discrete-phase model (DPM):** In the DPM models, the base phase is modeled using the continuum equations of fluids, and the dispersed phase is simulated by tracking fictitious solid particles or vapor/gas bubbles. The collision of particles with walls, the particle–particle, and the particle-base fluid interactions, the breakup and collision of bubbles and drops should be separately modeled. This method usually generates acceptable results for low-volume fraction dispersed phase or dilute multiphase flows. In Eulerian–Lagrangian trajectory models, a tracing equation for the dispersed phase is introduced based on the second law of Newton and flow kinematics,

$$\frac{d\mathbf{u}_i}{dt} = \mathbf{F}_{Di}, \quad \frac{d\mathbf{x}_i}{dt} = \mathbf{u}_i \tag{6.119}$$

The interfacial force equals the sum of the modified Stokes drag force $\mathbf{F}_D = A(\mathbf{u} - \mathbf{u}_i)$, the buoyancy force $\mathbf{F}_B = \mathbf{g}\frac{\rho_D - \rho}{\rho_D}$, the virtual mass or the added mass force $\mathbf{F}_{Added} = 0.5\frac{\rho}{\rho_D}\frac{d}{dt}(\mathbf{u} - \mathbf{u}_i)$, the thermophoretic force

$\mathbf{F}_{TP} = -D_{TP}\frac{1}{m_D T}\nabla T$, the Brownian force for small particles based on the Gaussian white noise stochastic term, and the Saffman's lift force. All such forces change the trajectory of particles, bubbles, or drops. $A = \frac{18\mu}{rho_D d_D^2}\frac{ReC_{Drag}}{24}$ is a coefficient in which C_D is a function of the particle Reynolds number defined based on the particle diameter d_D and $|\mathbf{u} - \mathbf{u}_i|$. There are several correlations in literature for $C_{Drag} = f(Re)$.

The FCM model: In the force coupling method (FCM), a point-force source term is added to the Navier–Stokes equation in the form [127]

$$F_{FCi} = -\sum_{n}^{0} F_i^n \delta(x - y^n) \tag{6.120}$$

where y^n is the position of the center of mass of particles and δ is the Dirac delta function. The force exerted on each particle is

$$F_i^n = -(m_p^n - m_f)\frac{dU_i(y^n)}{dt} + F_i^B \tag{6.121}$$

where $F_i^B = (\rho_p - \rho_f)\forall_p^n g_i$ is the gravity body force and $\forall$ is the volume of particles.

3. **The phase-change process:** The vapor–liquid phase change process appears in applications like cavitation (or nanocavitation) and boiling, which are involved with (nanoscale) bubble/droplet growing or collapsing in gas–liquid two-phase flows. In discrete simulations, some new modifications should be added regarding the bubble dynamics, the variable non-spherical shape of bubbles, thermal effects, and the latent heat/radiation considerations in the energy equation [142]. In mixture models, a condensation term or a vapor generation term should be added to the mass transport of the vapor phase, when the local pressure is less than the vapor pressure of the liquid.

The Rayleigh–Plesset equation can be used to compute the size of spherical bubbles as a function of time

$$r\frac{d^2 r}{dt^2} + \frac{3}{2}\left(\frac{dr}{dt}\right)^2 + \frac{4v}{r}\frac{dr}{dt} + \frac{2\sigma}{\rho r} + \frac{\Delta p}{\rho} = 0 \tag{6.122}$$

where ρ and v are the density and the kinematic viscosity of the liquid, σ is the surface tension coefficient of the bubble–liquid medium, and Δp is the pressure difference between inside and outside of the bubble [140].

To model the solid–liquid phase change process[8] in the phase-change materials (PCM) or non-premixed combustion, the enthalpy method has been introduced. This method takes into account the exchange of latent heat during melting or solidification. The simplicity of this method originates from the elimination of the need to model the interface between the solid and the liquid phases. In the enthalpy method, a latent heat content is assumed for each cell as a function of temperature. The total volumetric enthalpy is defined as

$$H(T) = h(T) + \rho_l \phi L = \int_{T_m}^{T} \rho_i c_i dT + \rho_l \phi L \tag{6.123}$$

where $H, h, \rho_l, \phi, L, T_m$ are the total volumetric enthalpy, the sensible volumetric enthalpy, the density of the liquid phase-change material, the melt fraction, the latent heat of fusion, and the melting temperature of the liquid. Simply speaking, the melt function ϕ equals 0 and 1 for the solid ($T < T_m$) and the liquid phases ($T > T_m$), respectively, and something between 0 and 1 when $T = T_m$. Then, the energy equation is

$$\frac{\partial H}{\partial t} + \nabla \cdot (\mathbf{u}H) = \nabla (k_i \nabla T) + S \tag{6.124}$$

where k_i is the conductivity of phase i.

6.9 MICRO/NANOSCALE MECHANICS

To simulate micro/nanoscale flows, we may work within three frameworks

- Microscale approach: simulation of the behavior of atoms.
- Mesoscale approach: simulation of the behavior of a collection of atoms called particles using statistical tools, based on the Boltzmann equation.
- Macroscale approach: ignoring atoms and simulation of the medium as a continuum (Q6.7).

6.9.1 GASEOUS FLOWS

In this section, we extend the Navier–Stokes equation for high-Knudsen gaseous flows. When the length-scale of the problem is comparable to the size of the mean-free-path of molecules, the no-slip and the no-temperature jump conditions are not valid anymore. In addition, weak forces such as capillary forces become important.

The Knudsen number is defined as

$$Kn = \frac{\lambda}{L} \tag{6.125}$$

where L is the characteristic length scale. The von-Karman relation between three dimensionless parameters is

$$Kn = \frac{M}{Re} \sqrt{\frac{\gamma \pi}{2}} \tag{6.126}$$

There are alternative modeling approaches such as the Boltzmann equation, the Schrödinger equation, and the molecular dynamics to simulate high-Knudsen flows. Here, we are discussing the continuum models to simulate non-continuum flows. Different non-continuum flow regimes have been defined for small-length-scale gaseous flows,

1. **Continuum regime** $Kn < 0.001$: The Navier–Stokes equation can be used along with the no-velocity-slip and no-temperature jump conditions at solid boundaries and interfaces.

2. **Slip regime** $0.001 \leq Kn \leq 0.1$: The rarefication effects start to grow near walls by the formation of the Knudsen layer. The Navier–Stokes equation as a first-order Chapman–Enskog expansion of the Boltzmann equation along with the velocity-slip and jumped temperature condition on walls can be used in this regime. The compressible form of the Navier–Stokes equations should be used with Maxwell's mixed-type boundary condition. The dimensionless form of the velocity-slip and the temperature-jump conditions are as follows.

$$
\begin{aligned}
u_f^* - u_w^* &= \frac{2 - \sigma_v}{\sigma_v} Kn \frac{\partial u_s^*}{\partial n^*} + \frac{3}{2\pi} \frac{\gamma - 1}{\gamma} \frac{Kn^2 Re}{Ec} \frac{\partial T^*}{\partial t^*} \\
T_s^* - T_w^* &= \frac{2 - \sigma_T}{\sigma_T} \frac{2\gamma}{\gamma + 1} \frac{Kn}{Pr} \frac{\partial T_s^*}{\partial n^*}
\end{aligned} \tag{6.127}
$$

where u_f^* and T_f^* are the slip-velocity and the jumped temperature of the gas over walls, respectively. σ_v and σ_T are the momentum and thermal accommodation coefficients, respectively. $\sigma_T = 1$ corresponds to the case with perfect energy exchange. $\sigma_v = 0$ and 1 represent the specular and diffuse reflections, respectively.

Temperature enters the velocity-slip condition via the second term as a tangential temperature gradient corresponding to thermal creep or transpiration phenomenon. When a rarified gas flow is subjected to a tangential temperature gradient along the walls, the gas starts to creep from the cold reservoir toward the hot one.

3. **Transitional regime** $0.1 < Kn < 10$: In addition to employing high-order slip and jumped conditions on walls, the Stokes constitutive relations should also be modified in the transitional regime. The second-order expansion of the Boltzmann equation valid for the early stage of this regime is called the Burnett equation. The Burnett equation can be regarded as the modified form of the constitutive relations at high-Knudsen numbers. Neglecting the thermal creep terms, the corrected boundary conditions are

$$
\begin{aligned}
u_f^* - u_w^* &= \frac{2 - \sigma_v}{\sigma_v} \left[Kn \frac{\partial u_s^*}{\partial n^*} + \frac{Kn^2}{2} \frac{\partial^2 u_s^*}{\partial n^*} \right] \\
T_s^* - T_w^* &= \frac{2 - \sigma_T}{\sigma_T} \frac{2\gamma}{\gamma + 1} \left[\frac{Kn}{Pr} \frac{\partial T_s^*}{\partial n^*} + \frac{Kn^2}{2Pr} \frac{\partial^2 T_s^*}{\partial n^{*2}} \right]
\end{aligned} \tag{6.128}
$$

4. **Free-molecular regime** $10 \leq Kn$ or near-vacuum flows. In such flows the gas is highly rarified, the pressure is negligibly small, and the collision term in the Boltzmann equation can be neglected.

Nanoscale Stokes flow: If the flow over a small particle is under consideration, the Conningham correction factor has to be used to compute the drag force exerted on the small particles. The Conningham factor corrects the Stokes drag force obtained from the creeping flow theory due to the slip condition over the surface of the particle at high Knudsen numbers. The drag coefficient obtained from the Stokes law is divided by the Conningham correction factor,

$$C_{D,corrected} = \frac{C_{D,Stokes}}{C} \tag{6.129}$$

where $C = 1 + 2Kn[A_1 + A_2\exp(-A_3Kn)]$ in which $A_1 = 1.275$, $A_2 = 0.400$, $A_3 = 0.55$ are experimentally determined constants for air. The Conningham factor is important for particles smaller than 15 micron and becomes inaccurate for particles smaller than 1 microns due to the random Brownian motion effects [154].

Nanoscale Reynolds equation: If the moving fluid in a narrow gap is a gas, then the nanoscale size of the gap leads to the appearance of non-continuum behavior in the Reynolds lubrication equation. The modified version of the dimensionless Reynolds equations has been presented containing the thermal creep term and the bearing dimensionless number $B = \frac{6\mu UL}{p_o h_o^2}$ in reference [155].

Liquid nanoscale flows: For small-scale liquid flows, the medium remains continuum, but instead, some new phenomena become important such as electrokinetic effects, electroosmosis, electrophoresis, dielectrophoresis, thermo/electro capillary effects, and bubble dynamics [147].

Mesoscale modeling: Definition of fictitious particles helps us to compute the probability of their existence at a specific location and time and with a specific velocity. This strategy also avoids solving the Poisson equation for pressure and therefore decreases the computational cost. Similar to other Lagrangian models, the nonlinear advection terms vanish in the Boltzmann equation. In this approach, we have neglected atoms, but not in a way that continuum mechanics does.

Derivation of the Boltzmann transport equation is based on the definition of a distribution function. The distribution function is a seven-parameter function, which demonstrates the probability of the presence of a particle at a specific location, time, and with a specific velocity. The change in probability of a particle due to intermolecular collisions can be modeled by a function denoting the rate of collision [143, 144],

$$f\left(\mathbf{r}+\mathbf{c}dt, \mathbf{c}+\frac{\mathbf{F}}{m}dt, t+dt\right) - f(\mathbf{r},\mathbf{c},t) = \frac{\partial f}{\partial t}\bigg|_{Collision} dt \tag{6.130}$$

Here, f is the distribution function, $\mathbf{c}$, $\mathbf{r}$, m are the velocity, the position, and the mass of particles, respectively, and $\mathbf{F}$ is the external force. Dividing both sides by dt and letting $dt \to 0$,

$$\frac{Df}{Dt} = \frac{\partial f}{\partial t} + \frac{\partial f}{\partial \mathbf{r}}\cdot\frac{\partial \mathbf{r}}{\partial t} + \frac{\partial f}{\partial \mathbf{c}}\cdot\frac{\partial \mathbf{c}}{\partial t} = \Omega(f) \tag{6.131}$$

where the source term Ω is the collision operator. Using $\mathbf{c} = \frac{\partial \mathbf{r}}{\partial t}$ and $\frac{\mathbf{F}}{m} = \frac{\partial \mathbf{c}}{\partial t}$

$$\frac{\partial f}{\partial t} + \mathbf{c}\cdot\frac{\partial f}{\partial \mathbf{r}} + \frac{\mathbf{F}}{m}\cdot\frac{\partial f}{\partial \mathbf{c}} = \Omega(f) \tag{6.132}$$

Neglecting the external body force,

$$\frac{\partial f}{\partial t} + \mathbf{c}\cdot\frac{\partial f}{\partial \mathbf{r}} = \Omega(f) \tag{6.133}$$

Finally, the celebrated Boltzmann transport equation is obtained

$$\frac{Df}{Dt} = \Omega(f) \qquad (6.134)$$

The collision operator vanishes in a free-molecular fully ballistic regime in which particles experience long-term motion. This form is called the collisionless Boltzmann equation. Due to the integral form of the collision operator, the Boltzmann equation is an integrodifferential equation.

After performing the Chapman–Enskog expansion, the Euler and the Navier–Stokes equations can be derived from the Boltzmann equation as zeroth- and first-order terms, respectively. However, there are singularities in higher-order terms and the well-posedness of the solution needs extra discussion. Under the single-relaxation time (SRT) approximation or the Bhatnagar–Gross–Krook model, which is appropriate for the binary collisions of particles, the Boltzmann equation reads

$$\frac{\partial f}{\partial t} + v_i \frac{\partial f}{\partial x_i} + F_i \frac{\partial f}{\partial v_i} = \frac{f - f_0}{\tau_v} + g$$

$$f_0 = 4\pi \left(\frac{m}{2\pi kT} \right)^{\frac{3}{2}} c^2 e^{-\frac{mc^2}{2kT}} \qquad (6.135)$$

where f_0 is the local equilibrium distribution function or the Maxwell–Einstein distribution, v_i is the fluid particle velocity, τ_v is the particle's relaxation time, and g represents the body force. By increasing τ, the collision operator decreases and the equilibrium condition is recovered. The Boltzmann equation is too difficult to be figured out in general due to its seven-dimensionality and complexity of the collision integral. Another limitation for directly solving the Boltzmann transport equation is the curved boundary capturing.

We may obtain different physics by employing different equilibrium distributions. The quantum version of the Boltzmann equation exists in which the collision operator is computed based on quantum mechanics principles [145]. Macroscopic quantities including density and velocity can be obtained with the aid of the distribution function as follows.

$$\rho(x_i, t) = m \int f(x_i, v_i, t) dv_i$$

$$\rho u_i(x_i, t) = m \int v_i f(x_i, v_i, t) dv_i$$

However, the above-mentioned form of the Boltzmann equation results in an isothermal temperature field. Hence, another equation should be added containing a new parameter called the internal energy distribution function,

$$\frac{\partial g}{\partial t} + v_i \frac{\partial g}{\partial x_i} = \frac{g - g_0}{\tau_c} \qquad (6.136)$$

where τ_c is the internal energy relaxation time. Then, the internal energy can be computed from

$$\rho \varepsilon = \int g(x_i, v_i, t) dv_i \qquad (6.137)$$

and $g_0 = \varepsilon f_0$.

6.9.2 THERMAL TRANSPORT

In this section, we review important concepts in the field of nanoscale heat conduction.

- **Phonons** imply the vibrational energy of the lattice and are the main origin of thermal transport in solids. They have dual particle-wave nature. Due to the connections between atoms, the excitation of a group of atoms from their equilibrium state leads to the propagation of vibrational wave pockets through the material. The speed of motion of phonons is the sound speed or the phonon group velocity.
- **Phonon scattering:** Phonons may deviate from their straight path [146]. This is known as the scattering process, including the normal (N) and the resistive (R) types,

  - Umklapp (U) phonon–phonon scattering when two or more phonons touch each other.
  - Phonon–imperfection scattering when phonons interact with impurities in the material.
  - Boundary-phonon scattering.
  - Phonon–electron scattering when the material is heavily doped.

- **Relaxation time:** the time needed to recover the equilibrium condition after the scattering process occurs.
- **Thermal Knudsen number** is defined similar to the hydrodynamic Knudsen number

$$Kn_T = \frac{\lambda}{L} = \frac{3k}{CvL}, \qquad (6.138)$$

  where λ and L are the mean-free-path of phonons and the characteristic length, respectively. k is the thermal conductivity, C is the specific heat per unit volume, and v is the average group velocity of phonons. If the thermal Knudsen number increases from zero to higher values, the non-Fourier models, the jumped-temperature at boundaries, the Boltzmann transport equation, the molecular dynamics modeling, and the Schrödinger equation should be used to predict the thermal behavior of matter.

  Based on the Knudsen number, the fully ballistic and the fully diffusive regimes are two limits of heat transfer corresponding to high-Knudsen and low-Knudsen problems, respectively. The pure diffusive behavior can be modeled by the Fourier law that exhibits the heat transfer paradox (infinite velocity of heat carriers).

- **Thermal metamaterials** are artificially manufactured materials with special thermal characteristics. Some thermal metamaterials are

  - Thermal cloak: eliminates the thermal gradient in a specific region.
  - Thermal diode or thermal rectifier: blocks the flow of heat in a specific direction.
  - Thermal rotator: changes the direction of the temperature gradient.

- Thermal concentrator: maps the heat transfer from a larger zone to a smaller one.
- Thermal camouflage: hides the temperature gradient in a region.

- **The phonon-Boltzmann transport equation** under the single model relaxation time (SMRT) approximation is

$$\frac{\partial f_\omega}{\partial t} + \vec{v}_\omega \cdot \nabla f_\omega = -\frac{f_\omega - f_\omega^0}{\tau_\omega} + g_{e-ph}, \qquad (6.139)$$

where f is the distribution function, f^0 is the equilibrium distribution function of Bose–Einstein, $\vec{v}$ is the heat-carrier group velocity, ω is the angular frequency of the heat carriers, and τ is the heat-carrier relaxation time. The term g_{e-ph} is representative of the phonon generation rate due to the electron–phonon scattering [148]. The equilibrium Bose–Einstein distribution reads

$$f_\omega^0 = \left(e^{\frac{\hbar\omega}{k_b T}} - 1 \right)^{-1} \qquad (6.140)$$

where $\hbar$ is the Plank's constant divided by 2π and k_B is the Boltzmann's constant. Integrating the Boltzmann equation over frequency, the energy density of phonons is

$$e(T) = \int f_\omega \hbar\omega D(\omega) d\omega \qquad (6.141)$$

where $D(\omega)$ is the phonon density of state per unit volume. The gray model assumes that all phonons have the same group velocity and relaxation time. Then, the subscript ω can be eliminated. Therefore, the energy density form of the Boltzmann equation under the gray assumption is

$$\frac{\partial e}{\partial t} + \vec{v} \cdot \nabla e = -\frac{e - e^0}{\tau} + q \qquad (6.142)$$

where e^0 and q are the equilibrium phonon energy density and the internal heat generation rate, respectively.

There are many other concepts in this framework such as coherence, phonon tunneling, phonon engineering, the Brillouin zone, phonon–photon interaction, relaxon, the phonon hydrodynamics regime, the Kapitza resistance, the Levy flight and the fractal super-diffusion of heat, the Monte Carlo method for phonons, and the lattice Boltzmann method for phonons. A review of terminologies and modeling approaches in different levels has been presented in reference [156].

6.9.3 MICROSCALE MODELING

Molecular dynamics uses Newton's second law of motion for each molecule

$$\mathbf{F}_i = m_i \frac{d^2 \mathbf{r}_i}{dt^2} \qquad (6.143)$$

where m_i is the mass. The exerted force on each atom is calculated using the intermolecular potential. The most famous potential function is the Lennard–Jones Potential

$$V(r) = 4\varepsilon \left[\left(\frac{\sigma}{r} \right)^{12} - \left(\frac{\sigma}{r} \right)^{6} \right] \tag{6.144}$$

The force is obtained by the relation $\mathbf{F} = -\nabla_r V(r)$ and σ is the hard sphere diameter. Other potentials are the Square-well, Yukawa, WCA, Buckingham, and the Coulomb potentials [147].

Quantum hydrodynamics: The Schrödinger equation is the fundamental partial differential equation of quantum mechanics, similar to the role of the Navier–Stokes equation in continuum fluid mechanics. In this section, we are going to present fluid dynamics aspects of the Schrödinger equation that governs the evolution of a wave function over time,

$$i\hbar \frac{\partial \psi}{\partial t} + \frac{\hbar^2}{2m} \nabla^2 \psi - V\psi = 0 \tag{6.145}$$

where $i = \sqrt{-1}$, $\hbar$ is the reduced Planck's constant, ψ is the wave function of a particle with mass m subjected to a potential V. The mathematical form of the equation is similar to the heat equation except the imaginary factor on the left-hand side.

The quantum thermo-fluid mechanics can present justifications for unusual effects such as superfluidity, the nanoscale intermolecular forces, quantum turbulence at near-zero temperatures, quantum vortex, quantum dissolving, and the quantized circulation as the quantized version of the classical circulation. The quantized circulation along any closed curve is $\Gamma = 2\pi n\hbar$. Quantum behavior of fluids is seen in cryogenic fluids, neutron stars, the quark-plasma phase of matter, and the Rollin film which exhibits creeping escape from a container related to the third sound phenomenon. The second sound is a quantum thermal phenomenon in superfluids or dielectric crystals. The second sound is related to the speed of the wave-like motion of phonons as heat carriers [151].

It is possible to derive the velocity vector from the wave function. This way, the Madelung equation as an alternative form of the Schrödinger equation will be obtained based on hydrodynamic variables as a quantum counterpart of the Navier–Stokes equation for the j-th fluid component,

$$\frac{\partial \rho_j}{\partial t} + \nabla \cdot (\rho_j \mathbf{u}_j) = 0$$

$$\frac{\partial (\rho_j \mathbf{u}_j)}{\partial t} + \nabla (\rho_j \mathbf{u}_j \mathbf{u}_j) = -\rho_j \nabla W_j - \rho_j \nabla v_{cou} - \rho_j \nabla v_{xc} \tag{6.146}$$

where $W_j = \frac{-0.5\nabla^2 \psi_j}{\psi_j}$ is the Bohm's quantum potential. The first to the third terms on the right-hand side denote the Bohm force density, the Coulomb force density, and the exchange-correlation force density. Equations (6.146) can be written in terms of

ρ and $\mathbf{u}$

$$\frac{\partial \rho}{\partial t} \quad + \quad \nabla \cdot (\rho \mathbf{u}) = 0$$

$$\frac{\partial}{\partial t}(\rho \mathbf{u}) \quad = \quad \nabla \cdot \mathbf{T} - \rho \nabla v_{cou} \qquad (6.147)$$

in which the stress tensor $\mathbf{T}$ is composed of the kinetic contribution of the electron fluid the thermal stress tensor, the Bohm stress tensor, and the exchange-correlation stress tensor. v_{cou} is the interpretation of chemical binding [152].

Another quantum equation, which is useful to describe the behavior of superfluids and quantum turbulence is the Gross–Pitaevskii equation [153]. In this context, the analogy between the Navier–Stokes equation and the Gross–Pitaevskii equation can be presented, and the quantum pressure as a result of the Heisenberg uncertainty principle and the mass density, $\rho = m|\psi|^2$, are introduced. m is the mass of the boson.

6.10 NON-CONVENTIONAL THERMO-FLUID MECHANICS

6.10.1 FRACTIONAL MECHANICS

When we have a non-integer derivative in an equation, the fractional or Leibniz's calculus should be used to define such a strange operator. In this framework, the partial derivative with respect to time or space is written in the fractional form. Physically, the non-integer fractional derivatives show non-local behavior, which may be meaningful in various applications like memory-effects, contaminant transport in porous media, non-Newtonian fluids, polymeric flows, acoustical wave propagation in biological media, nanotechnology, viscoelasticity, groundwater flows, and even in turbulence modeling. Such non-integer fractional derivatives can fill the mathematical gap between integer-order derivatives as a middle-ground.

One of the most applicable and well-known definitions of the fractional derivative is Caputo's relation

$$D_x^\alpha(f(x)) \quad = \quad \frac{d^\alpha f(x)}{dx^\alpha} = \frac{1}{\Gamma(n-\alpha)} \frac{d^n}{dx^n} \int_0^x (x-t)^{n-\alpha-1} f(t)dt, \quad n-1 < \alpha < n$$

$$D_x^\alpha(f(x)) \quad = \quad \frac{d^n f(x)}{dx^n}, \quad \alpha = n \qquad (6.148)$$

where Γ is the gamma function with the famous property $\Gamma(n+1) = n!$, α is the order of differentiation. One of the most important characteristics of Caputo's definition is that the Caputo fractional derivative of a constant is zero, and the fractional Taylor series may be written similar to the integer-order Taylor series [157].

1. **Fractional continuity equation:** The fractional conservation of mass equation [158] for incompressible flow is written based on the fractional divergence

operator,

$$\nabla^\alpha \cdot \mathbf{u} \equiv \frac{\partial^\alpha u_i}{\partial x_i^\alpha} = 0 \tag{6.149}$$

2. **The fractional time-derivative term** for the acceleration term in the Navier–Stokes equation [159]

$$\frac{D^\alpha \mathbf{u}}{Dt^\alpha} = \frac{\partial^\alpha \mathbf{u}}{\partial t^\alpha} + (\mathbf{u}_r \cdot \nabla)\mathbf{u} \tag{6.150}$$

3. **The fractional diffusion term** for the Laplacian term [160]

$$\frac{\partial^\alpha \mathbf{u}}{\partial t^\alpha} = -\nu(-\nabla^2)^\beta \mathbf{u} \tag{6.151}$$

where constants α and β correspond to the long-term decay and the diffusion non-locality, respectively. If these constants are functions of time and space, the equation is called the variable-order fractional equation. We have different cases, the typical diffusion $\beta = 1$, the super-diffusion $\beta > 1$, and the sub-diffusion $\beta < 1$. The non-local process is seen in biological tissues and porous media.

4. **The Levy flight and random walk in thermal transport:** The Levy flight illustrates the fractal super-diffusion of heat. The normal diffusion of heat generates a process governed by the Brownian motion of phonons. The Brownian motion in any stochastic system is the random and irregular movement of particles. The Levy dynamics describes a quasi-ballistic motion defined based on the random walk procedure and generalizes the Brownian motion by means of the fractal dimension parameter (α). α varies between 1 and 2 corresponding to pure ballistic and pure Fourier diffusion processes, respectively.

   In systems involving small length-scales relative to the phonon mean-free-path, the normal diffusion of heat originating from the Brownian motion starts to fail. The generalized governing differential equation to obtain the temperature distribution based on the Levy flight is

$$\frac{\partial T}{\partial t} = D\frac{\partial^\alpha T}{\partial x^\alpha} \tag{6.152}$$

   where D is the thermal diffusivity, and α denotes the dimension of the Riesz fractional space, which equals 2 and 1.6 for pure Brownian motion and the Levy flight, respectively. The self-affinity nature of the Levy process leads to the generation of fractal structures in nanoscale thermal transport [89].

5. **Fractional acoustics:** The fractional Zener wave equation is

$$\nabla^2 u - \frac{1}{c_0^2}\frac{\partial^2 u}{\partial t^2} + \tau_\sigma^\alpha \frac{\partial^\alpha}{\partial t^\alpha}\nabla^2 u - \frac{\tau_\varepsilon^\beta}{c_0^2}\frac{\partial^{\beta+2} u}{\partial t^{\beta+2}} = 0 \tag{6.153}$$

   where c_0 is the phase velocity at zero frequency, u is the acoustic variable (displacement in vibroacoustics), τ_σ and τ_ε are time constants referring to the

stress and strain, respectively. Also, a five-parameter Zener fractional constitutive relation for non-local elasticity has been defined as

$$\sigma(t) + \tau_\varepsilon^\beta \frac{\partial^\beta \sigma(t)}{\partial t^\beta} = E\left(\varepsilon(t) + \tau_\sigma^\alpha \frac{\partial^\alpha \varepsilon(t)}{\partial t^\alpha}\right) \tag{6.154}$$

where σ and ε are the stress and strain, respectively [161].

6. **Fractional quantum fluid mechanics:** The ordinary Schrödinger equation as
the governing equation of quantum mechanics was introduced in Section 6.9.3.
Here, we are referring to the variable-order fractional form of the Schrödinger
equation as follows.

$$i\hbar \frac{\partial^\alpha \psi}{\partial t^\alpha} - (-\frac{\hbar^2}{2m}\nabla^2)^{\beta/2}\psi - V\psi = 0 \tag{6.155}$$

α and β are the position-dependent levy constants [162].

6.10.2 STOCHASTIC MECHANICS

1. **Chaotic versus stochastic processes:** In various physical sciences, we face
chaotic and stochastic processes. Chaos appears in highly nonlinear systems,
which is super sensitive to even small changes in initial conditions. The behavior of such systems are somehow unpredictable due to the existence of numerous input parameters and drastically responding to initial conditions. So,
they are called chaotic. The most famous example is turbulent flow. A simple and inspiring equation to describe the chaotic behavior is the logistic map,
$x_n = \mu x_{n-1}(1 - x_{n-1})$, where $0 < \mu < 4$ and x_0 is an initial guess between 0
and 1 [163].

On the other hand, stochastic processes are related to randomness. Scientific
elements that contain a level of randomness, and their behavior is again unpredictable are stochastic. Well-known examples are the Brownian motion,
the Levy flight, the Wiener process, the random walk, the Markov chain, and
variation of the asset price. Unlike chaotic systems, the stochastic process is
insensitive to the initial conditions. The main difference between the chaotic
and the stochastic processes is that the results of a chaotic process are theoretically predetermined or deterministic but practically unpredictable.

All physical systems may contain a level of chaos. This characteristic is dom-
inant in chaotic systems. A question may arise here: does a stochastic process
really exist in nature? Albert Einstein says: "God does not play dice". Imagine
that you are tossing a coin. Is this process random? It is sophisticated but possible to compute the result of tossing if you solve equations of motion of the
coin by taking into account the drag force exerted from the air and the initial
velocity of the coin. So, tossing a coin is potentially deterministic, but it is
chaotic in the real world. Hence, we may investigate the problem by means of
statistical concepts.

However, the governing equations, the solution techniques, or the geometry can be stochastic or deterministic. If the equation governing a physical system contains a random-value function or if the numerical solution technique used to solve the equation contains a random-number generation, they are called stochastic. Deterministic equations such as the ordinary heat equation and the Navier–Stokes equations can be solved using stochastic methods like the Monte Carlo technique. Also, stochastic geometry of some materials like porous media called the random media may be reconstructed using randomly distributed artificial pores.

2. **Stochastic calculus:** The ordinary calculus of integration and differentiation for deterministic problems can be extended to stochastic systems by adding randomness using the white noise term. From a mathematical viewpoint, variables in stochastic calculus are fractals. This means that they are self-similar, and when you are zooming on the curves, they repeat themselves. But when you closely focus on deterministic curves, you will just see a straight line.

So, stochastic functions are not smooth or continuous functions, and the ordinary definitions in differential calculus like the chain rule cannot be used for such functions anymore. As a result, relations like $dX(t)$ and $\int dX(t)$ are allowed in stochastic calculus, but $\frac{\partial X}{\partial t}$ and $\int \frac{\partial X}{\partial t} dt$ are meaningless. The Itô calculus is the extension of the ordinary calculus, which presents definitions for stochastic calculus like the Itô chain rule.

Due to the randomness of the stochastic problems, we have to use concepts of statistics like ergodicity, probability density function (PDF), the Gaussian and normal distributions, the central limit theorem, the statistical moments such as mean (the first moment), variance (the second moment), skewness (the third moment), kurtosis (the fourth moment), hyperskewness (the fifth moment), and hypertailedness (the sixth moment). These concepts may help us to describe the nature of stochastic systems. Partial differential equations containing the random white noise term are called stochastic partial differential equations (SPDE). Based on the central limit theorem, the white noise term is a zero-mean Gaussian distribution.

3. **Mathematical modeling:** Consider a deterministic distribution $X(t) = X_0 e^{\mu t}$. μ controls the rate of increase of the curve starting from X_0. $\mu = 0$ represents the uniform distribution. Taking the derivative with respect to t and simplifying the result, we obtain

$$dX(t) = \mu X(t) dt \qquad (6.156)$$

Now, let us make the distribution stochastic, by adding the noise term W_t to the exponent $X(t) = X_0 e^{\mu t + \sigma W_t}$. The parameter σ controls the magnitude of fluctuations around the mean value. Again take the derivative and replace the aforementioned definition

$$dX(t) = \mu X(t) dt + \sigma X_t dW_t \qquad (6.157)$$

The second term in the right-hand side of the equation is the origin of the stochastic nature of the problem. Pay attention that the mean value of the noise term is zero, and as a result, the mean variation of the stochastic curve is the same as the solution of the noiseless equation. A general stochastic process can be described by

$$dX(t) = \mu(X(t),t)dt + \sigma(X(t),t)dW_t \tag{6.158}$$

where $X(t)$, μ, σ, and W_t are the stochastic parameter, the drift coefficient, the diffusion coefficient, and the standard Brownian or the Wiener zero-mean random fluctuations, respectively. The first and the second terms in the right-hand side of the equation are the drift and the Brownian white noise (the Wiener, the diffusion, the volatility, the shock) terms, respectively. Integrating Equation (6.158) from time 0 to T, yields

$$X(T) - X(0) = \int_0^T \mu(X(u),u)du + \int_0^T \sigma(X(u),u)dW_u \tag{6.159}$$

where u is the dummy variable of integration. Equation (6.158) can be simplified by assuming coefficients to be constants

$$dX(t) = \mu dt + \sigma dW_t \tag{6.160}$$

Equation (6.160) describes the arithmetic Brownian motion. Similarly, integrate the equation from 0 to T

$$X(T) - X(0) = \mu T + \sigma \int_0^T dW_t \tag{6.161}$$

4. **Mathematical finance:** Another form of Equation (6.160) can be derived by replacing $\mu[X(t),t] = X(t)\mu[X(t)]$ and $\sigma[X(t),t] = X(t)\sigma[X(t)]$

$$
\begin{aligned}
dX(t) &= \mu[X(t)]X(t)dt + \sigma[X(t)]X(t)dW_t \rightarrow \\
\frac{dX(t)}{X(t)} &= \mu[X(t)]dt + \sigma[X(t)]dW_t
\end{aligned}
\tag{6.162}
$$

which are called the geometric Brownian motion and are used in quantitative finance. Here, $X(t)$ is the price of the stock as a function of time. Discretizing the differential term $dX(t) = X(t + \Delta t) - X(t)$ indicates that the price at the next time-step just depends on the price at the previous step. This is the property of all Markov chain-type processes. Integrate the equation from 0 to T, using the Itô chain rule

$$
\begin{aligned}
\int d\ln X(t) &= \left(\mu - \frac{1}{2}\sigma^2\right)T + \sigma \int_0^T dW_t \rightarrow \\
\ln X(T) - \ln X(0) &= \left(\mu - \frac{1}{2}\sigma^2\right)T + \sigma \int_0^T dW_t \rightarrow \\
\frac{X_T}{X_0} &= e^{\left(\mu - \frac{1}{2}\sigma^2\right)T + \sigma \int_0^T dW_t}
\end{aligned}
\tag{6.163}
$$

Such stochastic differential equations can be solved using stochastic methods like the Monte Carlo or the Las Vegas methods based on random number generation methods (like the Box–Muller algorithms). It should be mentioned that there is another approach in mathematical finance based on differential equations called the Black–Scholes PDE approach, which is out of the scope of the present book.

5. **Stochastic thermo-fluid sciences:**

- One of the basic stochastic partial differential equations in thermo-fluid sciences is the Kardar-Parisi-Zhang equation [164],

$$\frac{\partial h(\mathbf{x},t)}{\partial t} = v\nabla^2 h + \lambda (\nabla h)^2 + \eta(\mathbf{x},t) \tag{6.164}$$

The terms in Equation (6.164) are the time rate of change of h, the diffusion term, the nonlinear term, and the white Gaussian zero-mean space-time noise field ($\eta = \sigma \dot{W}$), respectively.

- Thermal fluctuations can be modeled by a white-noise-type term in boundary or initial conditions [165]

$$T_{f,i} = T_i + \psi(t) \tag{6.165}$$

where $\psi(t)$ is the white noise term. A stochastic Eulerian–Lagrangian method has been presented for the fluid-structure interaction problems [166]. In this model, the fluid flow is simulated using the Eulerian method, the elastic solid is simulated by the Lagrangian approach, and thermal fluctuations are formulated using stochastic fields.

- The stochastic version of Burgers' equation reads

$$\frac{\partial u}{\partial t} + u\frac{\partial u}{\partial x} = v\frac{\partial^2 u}{\partial x^2} - \lambda\frac{\partial W}{\partial x} \tag{6.166}$$

where λ in the last term controls the white noise term as a function of time and space [167].

- The compressible form of the stochastic Navier–Stokes equation reads [168]

$$0 = d\rho + \nabla \cdot (\rho\mathbf{u})dt$$
$$\sigma(\rho, \rho\mathbf{u})dW = d(\rho\mathbf{u}) + \left[\nabla p(\rho) - v\nabla^2\mathbf{u} + \nabla \cdot (\rho\mathbf{u} \otimes \mathbf{u})\right.$$
$$\left. - (\lambda + v)\nabla\nabla \cdot \mathbf{u}\right]dt \tag{6.167}$$

The Stokes hypothesis as well as a relation between pressure and density (like the isentropic relation) are needed to close equations. As mentioned before, the solution of such stochastic equations are not continuous and smooth but rough and fractal-like. This fact may be a source of the ill-posedness of the solution.

- The Cahn–Hilliard equation[9] has been introduced to describe the process of phase separation of a binary alloy or fluid when the temperature has been reduced from a value above the critical temperature. The stochastic version of the Cahn–Hilliard equation driven by the white noise term is

$$dc + \left[D\nabla^2 \left(-\gamma \nabla^2 c + f(c) \right) \right] dt = dW_t \qquad (6.168)$$

where $f(c)$ may be $c^3 - c$ and $c \in [-1, 1]$ is the ratio of two species [169].

- Another application of the stochastic concept in fluid mechanics is modeling the particle transport [170] in multiphase discrete simulations where the motion of particles is followed by considering the second law of Newton and the exerted forces. One such force for submicron particles is the Brownian force, which is modeled by the Gaussian white noise term.

6. **Stochastic electrodynamics** or fluctuating electrodynamics discusses the random version of the Maxwell equations, which tries to present a framework to connect the classical and the quantum electrodynamics, and uses the stochastic nature of classical equations to present an accurate description of atomic quantum physics [171].

7. **Brownian motion:** To find a mathematical framework to describe the stochastic Brownian motion of liquid/gas particles, we may propose the molecular-statistical particle-based method (Einstein–Smoluchowski relation) or use a stochastic differential equation (Langevin equation). The molecular approach uses a probability density function and the convolution integral along with the Stokes law to find a diffusion coefficient that is related to the mean squared distance of particles. It can be shown that the mean squared distance or the diffusion length is proportional to the square root of time.
   The Longevin equation of motion of particles includes the white noise term

$$m \frac{d\mathbf{u}}{dt} = -\frac{\mathbf{u}}{\mu_b} + \mathbf{W}(t)$$

where m is the particle's mass, $\mu_b = \frac{1}{6\pi \mu R}$ is the mobility of particles obtained based on the Stokes law, and R is the radius of particles.

6.10.3 RELATIVISTIC FLUID MECHANICS

In Euclidean space, the distance between two points can be computed using $\sqrt{dx^2 + dy^2 + dz^2}$. When you imagine time as a fourth dimension, the distance between two points in four dimensional Minkowski space will be $\sqrt{dx^2 + dy^2 + dz^2 + d\tau^2}$, where $\tau = ict$ is the time coordinate. So, the distance between two events is

$$ds = \sqrt{dx^2 + dy^2 + dz^2 - c^2 dt^2} \qquad (6.169)$$

where c is the speed of light in vacuum and $i = \sqrt{-1}$. Based on this extension, we can define the four-dimensional del operator or the quad operator or the four-gradient operator

$$
\begin{aligned}
\Box &= \frac{\partial}{\partial \tau}\mathbf{l} + \frac{\partial}{\partial x}\mathbf{i} + \frac{\partial}{\partial y}\mathbf{j} + \frac{\partial}{\partial z}\mathbf{k} \\
&= \frac{\partial}{\partial \tau}\mathbf{l} + \nabla \\
&= \frac{-i}{c}\frac{\partial}{\partial t}\mathbf{l} + \nabla
\end{aligned}
\tag{6.170}
$$

Similarly, let's upgrade the Galilean-invariant Laplacian operator to the fourth dimension to define the d'Alembert operator which is Lorentz invariant,

$$
\Box^2 = \frac{-1}{c^2}\frac{\partial^2}{\partial t^2} + \nabla^2
\tag{6.171}
$$

Heat conduction: The Fourier's law can be extended to the four-dimensional space using the four-gradient operator, including a complex part

$$
\mathbf{q} = -k\Box T = -k\nabla T + k\frac{i}{c}\frac{\partial T}{\partial t}\mathbf{l}
\tag{6.172}
$$

Then, the relativistic heat conduction equation is

$$
\frac{\partial T}{\partial t} = \alpha\Box^2 T = \frac{-\alpha}{c^2}\frac{\partial^2 T}{\partial t^2} + \alpha\nabla^2 T
\tag{6.173}
$$

The first and the second laws of thermodynamics in the relativistic form respectively are

$$
\rho c_p \frac{\partial T}{\partial t} + \Box \cdot \mathbf{q} = 0
\tag{6.174}
$$

and

$$
\rho\frac{\partial s}{\partial t} = -\Box \cdot \left(\frac{\mathbf{q}}{T}\right) - \frac{1}{T^2}\mathbf{q}\cdot\nabla T + \frac{S}{T}
\tag{6.175}
$$

There is a celebrated theorem, Noether's theorem, which implies that the conservation laws are time-, space-, and orientation-invariant. The Lorentz invariance indicates that the laws of physics are invariant under a transformation between two coordinate frames moving at constant velocity with respect to each other. Einstein's assumption refers to the speed of light which should be independent of the velocity of the observer [172].

The continuity: All previous governing laws in previous sections have been obtained for the non-relativistic limit of the relativistic fluid mechanics, denoted by the dimensionless optic Mach number $\frac{u_c^2}{c^2} \to 0$ where c is the speed of light. Here, we present the relativistic version of governing equations of motion for fluids. Based on a new flux vector defined as the 4-vector $\mathbf{j} = (c\rho, j_x, j_y, j_z)$, the continuity equation reads

$$
\nabla \cdot \mathbf{j} + \frac{\partial \rho}{\partial t} = \partial_\mu \mathbf{J}^\mu = 0
\tag{6.176}
$$

where $\partial_\mu = \frac{\partial}{\partial x^\mu}$ is the 4-gradient operator in tensor notation and μ is the index of space–time dimension.

The momentum equation: The Euler equation in the relativistic frame can be derived using the energy–momentum tensor [173]

$$T(\mathbf{u}) = p \begin{pmatrix} -1 & 0 \\ 0 & \delta_{ij} \end{pmatrix} + \frac{\varepsilon + p}{\sqrt{1 - \mathbf{u}^2}} \begin{pmatrix} 1 & -u_j \\ -u_i & u_i u_j \end{pmatrix} \tag{6.177}$$

ε is the proper energy density. The energy–momentum tensor can be simplified for the fluid at rest by setting $\mathbf{u} = \mathbf{0}$. The relativistic Euler equation is

$$\frac{\partial \mathbf{u}}{\partial t} + \mathbf{u} \cdot \nabla \mathbf{u} = -\frac{1 - \mathbf{u}^2}{\varepsilon + p} \left(\nabla p + \mathbf{u} \frac{\partial p}{\partial t} \right) \tag{6.178}$$

and the compressible version

$$\frac{\partial \gamma \rho \mathbf{u}}{\partial t} + \nabla \cdot (\gamma \rho \mathbf{uu}) = -\frac{1}{\gamma f'(\rho)} \left(\nabla p + \mathbf{u} \frac{\partial p}{\partial t} \right) \tag{6.179}$$

where $\gamma = \frac{1}{\sqrt{1-\mathbf{u}^2}}$ and the function f represents the ε-density relation.

Perfect fluid is an ideal fluid that can be described just by the rest mass density and pressure. So, the energy–momentum tensor for perfect fluids is diagonal with the first component equal to the product of the rest frame mass density and the light speed squared [174]. The perfect fluids have zero shear stresses, viscosity, and heat conduction. A version of the Bernoulli equation in the relativistic context is

$$\frac{\partial}{\partial t} \left(\frac{\rho}{\sqrt{1 - \mathbf{u}^2}} \right) + \nabla \cdot \left(\frac{\rho \mathbf{u}}{\sqrt{1 - \mathbf{u}^2}} \right) = 0 \tag{6.180}$$

The energy equation is rewritten based on the concept of the rest mass [175]

$$\left[\frac{\partial \rho}{\partial t} + \nabla \cdot (\rho \mathbf{u}) \right] c^2 \;\; + \;\; \left[\frac{\partial}{\partial t} \left(\rho \left(\frac{u^2}{2} + h \right) \right) + \nabla \cdot \left(\rho \mathbf{u} \left(\frac{u^2}{2} + h \right) \right) \right]$$
$$+ \quad high - order\, terms = 0 \tag{6.181}$$

The first term is the rest mass part $O(c^2)$ and the second term is the flow energy part $O(u^2)$.

6.10.4 OTHER ASPECTS

Plasma flows: Plasma is the fourth state of matter. Generally speaking, plasma is a gas that contains charged particles like ions and free-electrons. Due to the high electrical conductance of plasmas, they can be affected by electromagnetic fields. Plasma may naturally be found in cosmological applications, stars, atmosphere, and can artificially be generated when a huge amount of energy is injected inside a gas like what happens in lightning, welding, and high-voltage circuit breakers. Plasmas are usually anisotropic and their properties differ along directions parallel to or perpendicular to the electromagnetic field.

Simulation of the motion of plasma can be studied in the framework of magnetohydrodynamics, which adds the Lorentz force to the equations of motion of fluids as a continuum. At the next level, the plasma may be treated as a two-fluid medium consisting of ions and electrons. One step forward, the behavior of plasma can be modeled using kinetic models such as the particle-in-cell or the Boltzmann equation with rather expensive computational cost [176].

Metafluid dynamics: Turbulence is probably the most sophisticated nonlinear phenomenon in nature. Metafluid dynamics is an interdisciplinary field of science that tries to create mathematical connections between diverse fields like turbulence, electrodynamics, and quantum mechanics [177].

Hemodynamics is a chaotic branch of fluid mechanics, which studies the blood flow in the circulatory vascular system and heart. Complexities exist such as elastic walls of the vessels, the blood-wall interactions, rheological behavior of blood (hemorheology), complicated geometry of the vessel network, the two-phase nature of blood due to the existence of cells, diffusion inside porous tissues, wall tension, local transition to turbulence, mass transfer, and metabolism [178].

Dark fluid acts as a bridge between two concepts of dark energy and dark matter in cosmology and astrophysics based on the general relativity theorem. The most interesting behavior of the dark fluid is its negative mass. Due to the extensive complexity of solution of equations in fluid mechanics, some simplified models like the logotropic fluid, the cosmic Chaplygin gas model, or the scalar field hypothesis have been presented to study such flows [179]. The relativistic version of the governing equations should be used to model the behavior of the dark fluid.

Machine/deep learning as a sub-field of artificial intelligence is an explosively growing branch of science that tries to teach machines. Machine learning can be connected to the fluid mechanics in applications like the boundary layer separation control and optimization [180], turbulence modeling [181], and low-Reynolds swimmers [182], drag reduction techniques [183], or using the computational fluid dynamics/experiments to present inverse mappings using deep learning.

Game theory is a mathematical framework to analyze the strategies when the success of a part of the system is dependent on the decisions of other parts. The game theory is linked to fluid mechanics via multi-objective stochastic control problems [184], the design optimization of internal aerodynamic shape operating at transonic flow [185], the periodic Burgers equation [184], and flow past a flat plate at zero incidence [186].

Virtual/augmented reality is a predetermined artificial world presented to a user via special equipment. If the content of such virtual space is generated using accurately validated CFD codes, the illustrated results will be truly real. So, we may construct the virtual classrooms, virtual labs, virtual wind tunnels with superpowerful capabilities for simulators, educational or marketing purposes [187].

Information theory is a mathematical theory including the process of information, communication, storage, transmission of information, and even entropy generation. Information technology is linked to fluid mechanics when we treat turbulence as information [188], in fluid-structure interaction problems [189], and entropic description of gravity [190].

3D computer graphics is linked to fluid mechanics via low-order numerical simulations of visible fluid motion using computer coding to generate detailed realistic high-quality animations or real-time computer games. The computational fluid dynamics tools, especially particle-based methods like the smoothed-particle-hydrodynamics or the level-set methods are appropriate choices for this purpose. Interesting elements of such computer graphic simulations may be the motion of water, air, foams, fire flames on the beach, melting of a monster, or even the motion of hair during a romantic scene [191].

NOTES

1. Change in surface energy and contact angle of a fluid interface in applications such as the capillary tube or drop dynamics.
2. Electrodynamics is a field of science that the predictions of quantum mechanics and special relativity come into agreement.
3. Other forces are the strong nuclear force, the weak nuclear force, and the gravitational force.
4. There is also a similar concept called the piezomagnetism. The name "piezo" comes from a Greek origin with the meaning "to push".
5. This form can be regarded as another alternative form of the Navier–Stokes equations.
6. "I am an old man now, and when I die and go to heaven there are two matters on which I hope for enlightenment. One is quantum electrodynamics, and the other is the turbulent motion of fluids. And about the former, I am rather optimistic". Horace Lamb (1932)
7. Like two types of solid particles of different sizes in a base fluid or two immiscible liquids in emulsions. Pay attention that a phase may consist of two or more substances.
8. The well-known benchmark problem in such applications is the Stefan problem, $\frac{\partial T}{\partial t} + u \frac{\partial T}{\partial x} = \alpha_l \frac{\partial^2 T}{\partial x^2}$ and $X(t) \leq x \leq \infty$.
9. Another equation in this category is the stochastic Allen–Cahn equation

7 Balance of Vorticity/Circulation Equation

In previous chapters, we already mentioned some points about the stream function-vorticity formulation of the Navier–Stokes equations. Then, we introduced the Crocco equation as a relation between vorticity and entropy. Despite the simple definition of vorticity vector[1] as the curl of the velocity vector, it is one of the most important parameters in analyzing and describing the behavior of turbulent and vortical flows. Because of the fluidity of fluids, some points in a flow field can move faster with respect to the neighboring points and vorticity (the molecular spin) appears.

On the other hand, we are seeking an equation governing the variation of vorticity called the vorticity equation. The vorticity equation helps readers gain a physical understanding of vortical flow patterns. It should be noted that, usually, we are not intended to directly solve the vorticity equation due to the complexities of applying boundary conditions. The vortex method is a Lagrangian mesh-free computational method, which tracks vortex particles based on the vorticity equation [192].

7.1 HOW TO MEASURE ROTATION? (Q7.1)

It is not possible to make a sensor to measure the magnitude of vorticity at a point. One of the challenging issues in fluid mechanics is to find a criterion to measure the magnitude of rotation and to investigate vortical regions in the flow. The most famous solution for this challenge is the vorticity vector, which equals twice the mean angular velocity of a fluid element. But, is the vorticity enough to demonstrate different aspects of vortical flows such as coherent structures in turbulent flows? This question motivates fluid mechanicians to find alternative techniques to measure rotation in fluids.

- **The vorticity vector:** From a physical viewpoint, the curl operator appears in the definition of the vorticity vector, which is proportional to the angular velocity of an element of a fluid. For a two-dimensional plane problem, the curl of a vector should be in the z-direction

$$\omega_z = \frac{\partial v}{\partial x} - \frac{\partial u}{\partial y}$$

Figure 7.1a illustrates a particle of fluid with different velocities at corners 1 to 4. Without loss of generality, suppose that the x-velocity of point 1 is smaller than that of point 2. So, $\frac{\partial u}{\partial y}$ is positive which leads to the rotation of

DOI: 10.1201/9781032719405-7

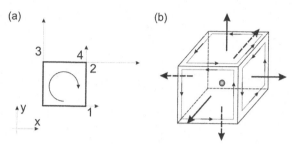

Figure 7.1 (a) A two-dimensional element to demonstrate the curl operator, (b) a cubic element with an overall output vorticity flux and positive non-zero divergence.

the element in the clockwise (or the negative z) direction. Similarly, if the vertical velocity of point 3 is greater than that of point 4, then $\frac{\partial v}{\partial x}$ becomes negative, which again leads to the rotation of the element in the clock-wise (or the negative z) direction. So, a negative sign should exist between two terms in the curl operator.

- **Circulation:** Using the curl theorem over an area A and its closed boundaries c,

$$\underbrace{\oint_c \mathbf{u}.\mathbf{dl}}_{Circulation} = \underbrace{\int_A \boldsymbol{\omega}.\mathbf{dA}}_{Vorticity \quad Flux} \quad \rightarrow \Gamma = \oint_c \mathbf{u}\cdot\mathbf{dl}$$

where **dl** is the differential element along the closed curve c. Equation (7.1) shows that the circulation along a closed curve is equal to the net flux of vorticity through the area inside the closed curve. If vorticity is zero at all points inside an area, the flux of vorticity crossing that area, and hence, circulation must be zero. The relation between the vorticity vector and circulation is somehow analogous to the relation between the velocity vector and the flow rate.

Characteristics of vorticity and circulation are listed here.

1. Dimensions of circulation and vorticity are L^2/t and $1/t$, respectively. Both are kinematic properties. Circulation is a scalar quantity, but the vorticity is a vector.
2. Circulation is a line-integral along a closed curve. Vorticity equals the curl of the velocity field at a point. So, circulation is a global integral parameter defined over a closed region. But, vorticity is a point function. Hence, circulation is a macroscopic parameter, but vorticity is a microscopic quantity.
3. Circulation determines the strength of a point-vortex as one of the elementary solutions of potential flows. Also, the well-known Kelvin theorem in vorticity dynamics discusses the condition under which the temporal variation of circulation is zero. In addition, the Kutta law in potential flow, states that the lift coefficient in flow over airfoils is proportional to the magnitude of the required circulation.

4. If the vorticity vector vanishes anywhere in a flow field, it is called irrotational. If the circulation is zero along any closed path inside the fluid, the flow is called acyclic. The reversed condition is cyclic motion. Rotational flows are essentially cyclic. Irrotational flows may be cyclic under some certain conditions [193].

• **Other approaches:** The following parameters or techniques may be useful to illustrate some different aspects of turbulent or vortical flows [194–196].

1. The helicity-based method. Helicity is proportional to the dot product of the velocity and the vorticity vector.
2. The second invariant of the velocity gradient tensor or the Q2-criterion, which can be helpful to detect some coherent structures.
3. The triple decomposition method,
4. The Δ-criterion, based on the eigenvectors and eigenvalues of the velocity gradient tensor,
5. The λ_2-criterion to find the pressure minimum,
6. The swirl strength parameters,
7. The predictor-corrector method,
8. The Lagrangian methods like the direct Lyapanov exponent and the M_z-criterion,
9. The R-definition, the pressure-minima, or the maximum vorticity method that tries to find the center of a vortex using the fact that the pressure is minimum at the eye of the vortex.
10. The combinatorial method,
11. Streamline-based tool to measure rotation,
12. The Rortex eigenvalue method.

7.2 VORTICITY EQUATION

7.2.1 REQUIRED IDENTITIES

Important vector identities that will be frequently used in this chapter are listed here.

1. Divergenceless vorticity

$$\nabla \cdot \boldsymbol{\omega} = 0 \qquad\qquad (Q7.2)$$

This is obvious since the divergence of the curl of any vector is zero. So, the vorticity field is a solenoidal vector field. This point implies that there should be similarities between the divergenceless velocity vector of incompressible flows and the vorticity vector. We define streamlines as curves that are tangent to the velocity vector, and a group of streamlines crossing a closed curve is called the stream tube. Similarly, the vortex line is defined as tangent to the local vorticity vector, and a group of vortex lines forms a vortex tube. Some well-known examples of the vortex tube are the vortex ring, smoke ring, and tornado.

The divergence-free property of vorticity implies that it is impossible to find a flow field with non-zero-divergence vorticity. Hence, based on the divergence theorem, the net flux of vorticity crossing any closed surface in space has to be zero,

$$\int_S \boldsymbol{\omega} \cdot d\mathbf{S} = \int_\forall \nabla \cdot \boldsymbol{\omega} d\forall = 0 \tag{7.1}$$

where S is a closed surface bounding the volume $\forall$. A similar result can be obtained for incompressible flows based on the conservation of mass. It is interesting to investigate this conclusion from a physical point of view. Consider that obtaining the divergence-free vorticity is not expected to contain any physical aspect, and is nothing more than a mathematical identity. Thus, our physical proof finally has to lead to a mathematical rationale anyway. As shown in Figure 7.1b, consider a point in space surrounded by an element of any shape, you say a cube. The divergence of a quantity at a point is the net sum of inputs and outputs to that point (the element in our analysis).

Use the reductio ad absurdum to prove the identity. Without loss of generality, consider an element with outward vorticity vectors on all faces, and hence, a nonzero (in our case) positive net flux of vorticity. So, the divergence of vorticity is positive. Based on the definition, the circulation along all surrounding perimeters of six faces of the cubic element (squares), and hence, the sum of circulations of six faces is a non-zero positive number. This is a contradiction since the overall sum of circulations of all faces surrounding a volume must be zero. The reason is that the total surface of the element is closed, and each side of the cube belongs to two neighboring faces. A part of the circulation of one side is canceled by the circulation of its neighboring face along the path they share in opposite directions.

Despite the similarity between velocity and vorticity (curl of velocity), there is an important difference between them. Vorticity is a local quantity but velocity has a global effect. From the physical point of view, a change in velocity anywhere affects the distribution of the velocity all over the domain. On contrary, the pattern of propagation of vorticity in the wake of a bluff body illustrates that the vorticity appears as separate plumes, which are gradually growing in size due to the diffusion of vorticity. So, we can sketch an approximate borderline between rotational and irrotational zones in a flow of a fluid. From a mathematical viewpoint, different behavior of velocity and vorticity stems from the pressure gradient term in the Navier–Stokes equation leading to an elliptic equation for pressure, Equation (4.33).

2. Curl of gradient

$$\nabla \times \nabla \phi = 0$$

This identity helps us to eliminate the gradient of any scalar quantity such as pressure or the velocity squared from equations. It implies that the pressure forces without the help of density change never can produce vorticity.

3. Decomposition of the advection term

$$(\mathbf{u} \cdot \nabla)\mathbf{u} = \nabla \left(\frac{1}{2}\mathbf{u} \cdot \mathbf{u} \right) - \mathbf{u} \times \boldsymbol{\omega}$$

This one is useful to deal with the nonlinear advection term in the material derivative operator. The last term in the equation is called the Lamb vector

4. Curl of the Lamb vector

$$
\begin{aligned}
\nabla \times (\mathbf{u} \times \boldsymbol{\omega}) &= -\boldsymbol{\omega}(\nabla \cdot \mathbf{u}) + (\boldsymbol{\omega} \cdot \nabla)\mathbf{u} - (\mathbf{u} \cdot \nabla)\boldsymbol{\omega} + \mathbf{u}(\nabla \cdot \boldsymbol{\omega}) \\
&= -\boldsymbol{\omega}(\nabla \cdot \mathbf{u}) + (\boldsymbol{\omega} \cdot \nabla)\mathbf{u} - (\mathbf{u} \cdot \nabla)\boldsymbol{\omega} \quad (7.2)
\end{aligned}
$$

The term $\mathbf{u}(\nabla \cdot \boldsymbol{\omega})$ is zero referring to the first identity.

5. The curl of the product of a scalar and the gradient of another scalar

$$
\begin{aligned}
\nabla \times (f\nabla g) &= \nabla f \times \nabla g + f\nabla \times \nabla g \\
&= \nabla f \times \nabla g \quad (7.3)
\end{aligned}
$$

The term $f\nabla \times \nabla g$ is zero due to the second identity. The proofs of these vector identities were presented in Section 1.4.

7.2.2 VORTICITY EQUATION

We trigger the derivation of the vorticity equation by expanding the material derivative term (using the third identity) and taking the curl of the Navier-Stokes equation.

$$
\frac{\partial \boldsymbol{\omega}}{\partial t} + \nabla \times \left[\nabla \left(\frac{1}{2}\mathbf{u}.\mathbf{u} \right) - \mathbf{u} \times \boldsymbol{\omega} + 2\boldsymbol{\Omega} \times \mathbf{u} \right] =
$$
$$
- \nabla \times \frac{\nabla p}{\rho} + \nabla \times \frac{\nabla \cdot \boldsymbol{\tau}}{\rho} + \nabla \times \frac{\mathbf{B}}{\rho} + \nabla \times \{[1 - \beta(T - T_0)]\mathbf{g}\} \quad (7.4)
$$

The curl of $\nabla \left(\frac{1}{2}\mathbf{u}.\mathbf{u} \right)$ is zero since the curl of gradient of any scalar field is null. The third term in the left-hand side (the Lamb vector) can be replaced using the fourth identity listed in the previous section.

$$
\frac{\partial \boldsymbol{\omega}}{\partial t} - [-\boldsymbol{\omega}(\nabla \cdot \mathbf{u}) + (\boldsymbol{\omega} \cdot \nabla)\mathbf{u} - (\mathbf{u} \cdot \nabla)\boldsymbol{\omega}] + \nabla \times (2\boldsymbol{\Omega} \times \mathbf{u})
$$
$$
= -\nabla \times \frac{\nabla p}{\rho} + \nabla \times \frac{\nabla \cdot \boldsymbol{\tau}}{\rho} + \nabla \times \frac{\mathbf{B}}{\rho} - \nabla \times [\beta T\mathbf{g}] \quad (7.5)
$$

The first and the fourth terms in the left-hand side are combined to create the material derivative of the vorticity vector. The curl of the pressure gradient term can be rewritten in the following form,

$$
-\nabla \times \frac{\nabla p}{\rho} = - \left[\nabla \left(\frac{1}{\rho} \right) \times \nabla p + \frac{1}{\rho}\nabla \times \nabla p \right] = - \left[-\frac{\nabla \rho}{\rho^2} \times \nabla p \right] = \frac{\nabla \rho \times \nabla p}{\rho^2} \quad (7.6)
$$

In the equation above, the curl of the gradient of pressure is zero. After rearranging other terms, the final form of the vorticity equation as an advection-diffusion equation, sometimes called the Helmholtz equation, reads

$$\underbrace{\frac{D\boldsymbol{\omega}}{Dt}}_{1} = \underbrace{(\boldsymbol{\omega} \cdot \nabla)\mathbf{u}}_{2} - \underbrace{(\nabla \cdot \mathbf{u})\boldsymbol{\omega}}_{3}$$

$$+ \underbrace{\frac{\nabla\rho \times \nabla p}{\rho^2}}_{4} + \underbrace{\nabla \times \frac{\nabla \cdot \boldsymbol{\tau}}{\rho}}_{5} + \underbrace{\nabla \times \frac{\mathbf{B}}{\rho}}_{6} - \underbrace{\nabla \times (2\boldsymbol{\Omega} \times \mathbf{u})}_{7}$$

$$+ \underbrace{\beta \mathbf{g} \times \nabla T}_{8} \tag{7.7}$$

The seventh term is just added by computing the curl of the Coriolis force where Ω is the angular velocity of a rotating frame such as the earth or the impeller of a pump. The Buoyancy term has been obtained using the vector identity $\nabla \times (T\mathbf{g}) = T\nabla \times \mathbf{g} + \nabla T \times \mathbf{g} = -\mathbf{g} \times \nabla T$. Explanations of eight terms in the vorticity equation are as follows (Q7.3).

1. **The material derivative** of vorticity indicates the total rate of variation of vorticity with time.
2. **The vortex stretching/turning** A vortex may stretch or twist under influence of the velocity gradient field. To clarify the physical interpretation of this term, consider the component of this vector quantity along the y-direction,

$$(\boldsymbol{\omega} \cdot \nabla)\mathbf{u}|_y = \underbrace{\omega_x \frac{\partial u_y}{\partial x}}_{Turning,Twisting,Tilting} + \overbrace{\omega_y \frac{\partial u_y}{\partial y}}^{Stretching,Squeezing} + \underbrace{\omega_z \frac{\partial u_y}{\partial z}}_{Turning,Twisting,Tilting}$$

$$\tag{7.8}$$

For a two-dimensional case inside the x–z plane, we have $\omega_x = \omega_z = \frac{\partial}{\partial y} = 0$. So, all three terms are zero for plane flows. This trend is expected because a two-dimensional flow should remain in the plane of motion and cannot twist (the first and the third terms are zero). In addition, the vortex stretching or squeezing along the direction normal to the flow plane is impossible due to the two-dimensionality of the flow field (the second term is also zero).

The first and the third terms on the right-hand side of Equation (7.8) are vortex turning, twisting, or tilting and the second term is the indicator of vortex stretching or squeezing. As shown in Figure 7.2a, if the velocity of point A is greater than the velocity of point B, based on the vortex stretching term, we conclude

$$V_A > V_B \rightarrow \frac{dV}{dy} > 0 \rightarrow (\boldsymbol{\omega} \cdot \nabla)\mathbf{u} = \omega\frac{dV}{dy} > 0 \rightarrow \omega\uparrow \tag{7.9}$$

This condition happens, for example, when a vortex propagates in a nozzle. Figure 7.2b illustrates a moving vortex inside a converging nozzle. Based

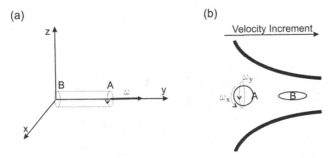

Figure 7.2 Configuration of vortex stretching in (a) a line-vortex along y-coordinate, (b) a vortex in a converging nozzle.

on the mass conservation law, it is obvious that the velocity increases in the streamwise direction. Referring to Equation (7.9), this trend leads to an increment of vorticity and stretching of vortex along the centerline. If we axially stretch a rotating object like what the mother of nature does in tornados, the angular velocity of that object shall be increased to preserve the angular momentum. Some well-known examples of the vortex stretching are the Burgers vortex, the Lamb–Oseen vortex, and the Hill vortex. Also, one of the mechanisms of dissipation in the energy cascade of turbulent flows is vortex stretching, which makes the vortex thinner and hence, easier to dissipate.

3. **Compressible vortex stretching** denotes stretching or squeezing of a vortex in compressible flows. Vorticity in this term appears as a multiplier of the velocity divergence that is representative of the density change. So, it can be concluded that this term is zero if the flow is irrotational or the velocity field is divergence-free. The negative sign behind this term indicates that if an element of fluid has positive angular velocity (vorticity) and the element expands, the vorticity will decrease and vice versa.

4. **The solenoidal term:** If the isodensity (isochore or isopycnic) lines and isopressure (isobar) lines do not coincide, the fluid element rotates in a way that these lines lie on each other. The fluid in which the isochore and the isobar lines are parallel is called the barotropic fluid, and the reversed condition is for the baroclinic fluid. Such vorticity production appears in atmospheric and oceanic flows.

Consider a fluid element with the distribution of density and pressure as shown in Figure 7.3. In this case, the density and the pressure are higher over the left side and the lower side of the element, respectively. So, the gradients of density and pressure are along the x and y directions, respectively. Based on the right-hand-rule, the cross product of the gradients of density and pressure ($\frac{\nabla \rho \times \nabla p}{\rho^2}$) is in +z direction.

This trend is expected based on two facts. First, the higher density of the fluid near the left side of the element indicates that the center of mass of the element should be closer to the left side of the element. Second, the direction of

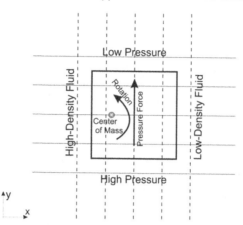

Figure 7.3 Distribution of pressure and density over a non-barotropic fluid element.

the pressure force is always from the high-pressure to the low-pressure zone (proportional to the negative of pressure gradient). So, the pressure force, in this case, is upward. These two comments together prove that the rotation of the element should be counter-clock-wise or parallel to the +z-direction.

5. **Diffusion of vorticity:** For an incompressible flow of an isotropic fluid with constant viscosity, the divergence of viscous stress tensor reduces to the Laplacian of the velocity field. So, the diffusion term can be written as

$$\nabla \times \frac{\nabla \cdot \tau}{\rho} = \nabla \times [\nu \nabla^2 \mathbf{u}] = \nu \nabla^2 \boldsymbol{\omega} \qquad (7.10)$$

It should be noted that the curl of Laplacian is equal to the Laplacian of curl. The mechanism of diffusion and convection of vorticity is similar to other transport processes such as diffusion of mass, heat, and momentum. If we only keep the material derivative term and the diffusion of vorticity in Equation (7.7), the vorticity equation reduces to

$$\frac{D\boldsymbol{\omega}}{Dt} = \nu \nabla^2 \boldsymbol{\omega} \qquad (7.11)$$

This is similar to the heat equation written for convective heat transfer. This fact indicates that the shedding of higher-vorticity cells behind a bluff body is similar to the motion of a painty fluid (a hot fluid) when the paint (the heat) is injected from the surface of the body. The physical result of the vorticity diffusion term is the growth of the size of vorticity bubbles inside the von Karman vortex street when they propagate downstream.

6. **The body force term** is related to the rotation of the fluid element due to the exertion of the body force. It should be noted that if the body force is conservative, the curl of the force is zero. Hence, this term vanishes in a constant-density flow when the body force is conservative.

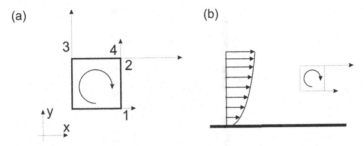

Figure 7.4 (a) Distribution of a non-conservative force over the surface of a fluid element, (b) the velocity profile and distribution of velocity on an element of fluid near the solid wall.

Figure 7.4a illustrates the distribution of a general two-dimensional non-conservative force over an element of fluid. The distribution of forces states that if the element rotates about the z-axis (in our case), the forces on points 1 to 4 should differ. It means that if $B_2 > B_1$ or $\frac{\partial B_x}{\partial y} > 0$ and $B_4 < B_3$ or $\frac{\partial B_y}{\partial x} < 0$, the element should start to rotate in the negative z-direction. This is exactly the z-component of the curl vector, which is $\frac{\partial B_y}{\partial x} - \frac{\partial B_x}{\partial y}$. The negative sign in this relation is consistent with our analysis.

7. **The Coriolis acceleration** can be simplified to

$$\nabla \times (2\mathbf{\Omega} \times \mathbf{u}) = 2\left[\mathbf{\Omega}\nabla \cdot \mathbf{u} - (\mathbf{\Omega}.\nabla)\mathbf{u} - \mathbf{u}(\nabla \cdot \mathbf{\Omega}) + (\mathbf{u} \cdot \nabla)\mathbf{\Omega}\right] \qquad (7.12)$$

The first term on the right-hand side is zero due to the incompressibility condition. The third and the last terms are zero for the case of the constant angular velocity of the rotating frame. So,

$$\nabla \times (2\mathbf{\Omega} \times \mathbf{u}) = -2(\mathbf{\Omega} \cdot \nabla)\mathbf{u} = -2\Omega_z\frac{\partial \mathbf{u}}{\partial z} \qquad (7.13)$$

The last equality is valid when the rotational speed is in the z-direction like the angular velocity of the earth, $\mathbf{\Omega} = \Omega_z\mathbf{k}$ [197]. Consider that the centripetal acceleration does not contribute to vorticity since it is a curl-free vector field.

8. **The buoyancy term:** The source term obtained from the Boussinesq approximation presents the effect of vorticity generation due to free convection. This term demonstrates that the vorticity has been generated normal to the gravity acceleration and the temperature-gradient vector. So, a temperature-gradient vector parallel to the gravity acceleration does not produce vorticity. Suppose that the gravity acceleration is in the negative vertical direction. Then, the component of the temperature-gradient vector in the horizontal plane generates a vorticity vector in the horizontal plane.

The direction of the increase of temperature in Figure 7.5 is assumed to be from left to write and the temperature gradient is positive in this case. The gravity vector is downward. So, the fluid starts to rotate in a counter-clockwise direction around the coordinate normal to the paper. This result is consistent with the cross product of the temperature gradient and the gravity vectors.

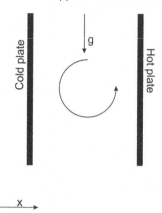

Figure 7.5   Vorticity generation due to the buoyancy force.

Curvilinear coordinates: Consider an axisymmetric flow in the r–z plane, in which the vorticity vector only has the following single component,

$$\boldsymbol{\omega} = \left(\frac{\partial u_r}{\partial z} - \frac{\partial u_z}{\partial r} \right) \mathbf{e}_\theta = \omega \mathbf{e}_\theta \tag{7.14}$$

Now, we can simplify the vorticity equation for inviscid incompressible flow ignoring the body forces and other sources of vorticity production,

$$\frac{D\boldsymbol{\omega}}{Dt} = \left[\frac{\partial \omega}{\partial t} + \left(u_r \frac{\partial}{\partial r} + u_z \frac{\partial}{\partial z} \right) \omega \right] \mathbf{e}_\theta$$
$$(\boldsymbol{\omega} \cdot \nabla)\mathbf{u} = \frac{\omega}{r} u_r \mathbf{e}_\theta \neq 0 \tag{7.15}$$

It is seen that the vortex shedding in two-dimensional axisymmetric flow is non-zero due to the non-zero derivative of the unit base vectors. Equating the second sides of equations (7.15), multiplying both sides by $\frac{1}{r}$, adding a zero term $\omega u_z \frac{\partial}{\partial z} \left(\frac{1}{r} \right)$, and using the rule of derivative of product, the vorticity equation reduces to

$$\frac{\partial}{\partial t} \left(\frac{\omega}{r} \right) + \left(u_r \frac{\partial}{\partial r} + u_z \frac{\partial}{\partial z} \right) \left(\frac{\omega}{r} \right) = \frac{D}{Dt} \left(\frac{\omega}{r} \right) = 0 \tag{7.16}$$

Sources of vorticity production: As presented in the previous part, the vorticity equation and the Crocco equation contain source terms that are the origins of the vorticity production. Pay attention that the vortex stretching terms and the vortex diffusion are not source terms of the equation and just can change the length and direction of the already produced vorticity by other terms. On the other hand, all differential equations need some boundary conditions and may have source terms. So, the production of vorticity inside the domain by the source terms and the generation of vorticity from boundaries and the incoming flow, should be considered.

1. Entropy/stagnation enthalpy gradients based on the Crocco theorem. The most famous evidence of this type of vorticity generation is in the region behind a curved shock wave in supersonic compressible flows. Since the strength of a shock wave depends on its angle and the angle of a curved shock wave is not constant, the entropy production along the curved shock wave is not uniform and an entropy gradient appears there.

2. Non-barotropicity of the fluid corresponding to the term number four in the vorticity equation.

3. Variation of density and exertion of a non-conservative body force such as the Lorentz force to an electrically conducting fluid under the influence of the magnetic field. This item is related to the term number six in the vorticity equation.

4. The non-inertial effects due to the Coriolis acceleration which becomes important in large-scale flows when the Rossby number is small. This term can be computed using the term number seven in the vorticity equation.

5. Natural convection, as discussed after the term number eight in Equation (7.7).

6. Walls. Referring to the analogy between the heat and the vorticity transports, similar to the injection of energy to the flow by imposing higher temperatures on walls, the wall injects vorticity to the flow. It means that the vorticity is born on walls and then convected, diffused, stretched, twisted, or squeezed inside the flow domain. So, even if all source terms of the vorticity equation are zero, the vorticity can be generated on solid boundaries of the domain. Rearrange the terms in the Navier–Stokes equation neglecting the body force over the no-slip permeable wall,

$$\left(\mu \frac{\partial \omega_z}{\partial y}\right)_w = -\left(\frac{\partial p}{\partial x}\right)_w - \left(\rho v \frac{\partial u}{\partial y}\right)_w \tag{7.17}$$

This relation proves that the permeability of the walls and the pressure gradient can generate the flux of vorticity through the wall. Without the suction or blowing and also with zero pressure gradient, the slope of vorticity normal to the wall is zero. Suction of fluid from walls and a negative pressure-gradient both lead to the creation of a positive slope of vorticity on the wall and vice versa. Pay attention that Equation (7.17) speaks about the variation (flux) of vorticity on walls, not the vorticity itself. As an example, both terms on the right-hand side are zero for the fully developed Couette flow, but the vorticity equals a non-zero constant in the entire domain. A similar discussion can be done for the flux of vorticity on walls in a Poiseuille-type flow.

Another physical description of the magnitude of vorticity may be fruitful. Based on the no-slip condition, the velocity of the fluid on walls relative to the wall is zero. So as shown in Figure 7.4b, a low-velocity region is created near walls which produces a velocity gradient across the element. So, for a two-dimensional case shown in Figure 7.4b, the vorticity appears in fluid along the negative direction normal to the page.

7. Vorticity of the upstream flow: If the incoming fluid carries a finite amount of vorticity, it will be convected, stretched, or diffused inside the domain. This item is not production of vorticity. It is a source of incoming vorticity.

7.3   CIRCULATION EQUATION

We use circulation as a macroscopic parameter directly related to the vorticity by the curl theorem. Consider a closed curved loop in space that is moving with a stream of fluid. To obtain the general form of Kelvin's theorem or the balance of circulation, we start by taking the material derivative of circulation along a curved loop,

$$\frac{D\Gamma_a}{Dt} = \frac{D}{Dt}\oint \mathbf{u}_t \cdot \mathbf{dl} = \oint \frac{D\mathbf{u}_t}{Dt} \cdot \mathbf{dl} + \underbrace{\oint \mathbf{u}_t \cdot \frac{D\mathbf{dl}}{Dt}}_{0} = \oint \frac{D\mathbf{u}_t}{Dt} \cdot \mathbf{dl} \qquad (7.18)$$

where $\mathbf{u}_t = \mathbf{u} + \mathbf{\Omega} \times \mathbf{r}$ and $\Gamma_a = \oint_c(\mathbf{u} + \mathbf{\Omega} \times \mathbf{r}) \cdot \mathbf{dl}$ are the absolute velocity and the absolute circulation, respectively. The material derivative of $\mathbf{dl}$ for a line element in the last term is simply $\mathbf{du}_t$. Then, the second term turns out to be zero:

$$\oint \mathbf{u}_t \cdot \frac{D\mathbf{dl}}{Dt} = \oint \mathbf{u}_t \cdot \mathbf{du}_t = \oint d\left(\frac{u_t^2}{2}\right) = \oint \nabla\left(\frac{u_t^2}{2}\right) \cdot \mathbf{dl} = 0 \qquad (7.19)$$

The line-integral of the gradient of any quantity along a closed curve is zero (the last integral) based on the Stokes integral theorem and the fact that the curl of the gradient of a quantity is zero.

Now, substitute the material derivative of velocity in Equation (7.18) using the Navier–Stokes equation. This way, a governing equation for circulation is obtained,

$$\frac{D\Gamma_a}{Dt} = \oint -\frac{\nabla p}{\rho} \cdot \mathbf{dl} + \oint \frac{[\mu\nabla^2\mathbf{u} + (\mu + \lambda)\nabla(\nabla \cdot \mathbf{u})]}{\rho} \cdot \mathbf{dl}$$
$$+ \oint \frac{\mathbf{B}}{\rho} \cdot \mathbf{dl} + \oint [1 - \beta(T - T_0)]\mathbf{g} \cdot \mathbf{dl} \qquad (7.20)$$

The Coriolis term is included in the definition of absolute circulation. This version of the equation is called the Poincare–Bjerknes circulation theorem.

Using the vector identity $\nabla^2\mathbf{u} = \nabla(\nabla \cdot \mathbf{u}) - \nabla \times \boldsymbol{\omega}$ and considering the fact that the closed-curve-integral of gradient of any quantity is zero,

$$\underbrace{\frac{D\Gamma_a}{Dt}}_{1} = \underbrace{\oint -\frac{\nabla p}{\rho} \cdot \mathbf{dl}}_{2} + \underbrace{\oint -\nu\nabla \times \boldsymbol{\omega} \cdot \mathbf{dl}}_{3}$$
$$+ \underbrace{\oint \frac{\mathbf{B}}{\rho} \cdot \mathbf{dl}}_{4} + \underbrace{\oint [1 - \beta(T - T_0)]\mathbf{g} \cdot \mathbf{dl}}_{5} \qquad (7.21)$$

1. The first term denotes the total time-rate of change of circulation along a moving closed curve.
2. The second term is the baroclinic term, which is important in baroclinic fluids where the lines of iso-density are not parallel to the lines of iso-pressure. This is the case of the rotation of air in the sea breeze. This term can be rewritten in a familiar form using the Stokes theorem,

$$\oint -\frac{\nabla p}{\rho} \cdot \mathbf{dl} = \int_A \frac{\nabla\rho \times \nabla p}{\rho^2} \cdot \mathbf{dA} \qquad (7.22)$$

3. The third term represents the circulation due to viscous forces and is connected to the diffusion of vorticity,

$$\oint v\nabla^2 \mathbf{u} \cdot \mathbf{dl} = \int_A v\nabla^2 \boldsymbol{\omega} \cdot \mathbf{dA} \tag{7.23}$$

4. The term number four denotes the circulation of non-conservative body forces. Pay attention that conservative forces do not produce circulation.
5. The term number five is the source of generation of circulation due to natural convection.

It is seen that the balance of circulation along a closed curve is equivalent to the balance of vorticity flux crossing the area confined to the closed path. On the other hand, we concluded that the circulation equation can be derived by taking the line integral of the dot product of the Navier–Stokes equation and the line element along a closed curve.

It should be noted that the vortex stretching terms (number 2 and 3 in Equation (7.7)) do not appear in the circulation balance. Equation (7.18) states that the material derivative of circulation equals the line-integral of the material derivative of the absolute velocity. This means that the material derivative and the line integration can be interchanged.

Theorems: Well-known theorems of vorticity dynamics help us to understand the behavior of vortical flows. They have been presented for inviscid flows, but are inspiring for viscous flows (Q7.4).

1. **The Kelvin circulation theorem:** In barotropic ideal flow in which the density is only a function of pressure, if the body force is conservative, circulation around any arbitrary closed curve remains constant during the advection of the vortex in the flow.

$$\frac{D\Gamma}{Dt} = 0 \tag{7.24}$$

This is the simplest version of the balance of circulation Equation (7.21) with all terms on the right-hand side being equal to zero.
2. **The Helmholtz first theorem:** Strength of a vortex filament and a vortex tube in inviscid flow remains constant along their length. The strength of a vortex reduces in realistic flows due to the dissipative effect of viscosity.
3. **The Helmholtz second theorem:** A vortex filament cannot end somewhere inside inviscid flows. It should form a closed loop or touch a solid boundary. Another statement of the theorem is that the fluid element on a vortex line remains on that line and the vortex line moves together with the flow.
4. **The Lagrange theorem or the Helmholtz third theorem:** When non-conservative body forces are absent, an initially irrotational inviscid flow will keep its irrotationality as time evolves. This theorem can be easily proved using the vorticity equation by assuming $\boldsymbol{\omega} = 0$,

$$\frac{D\boldsymbol{\omega}}{Dt} = (\boldsymbol{\omega} \cdot \nabla)\mathbf{u} \longrightarrow \frac{D\boldsymbol{\omega}}{Dt} = 0 \tag{7.25}$$

Other concepts and related topics are viscous potential flow [198], helicity $\int_\forall \mathbf{u} \cdot (\nabla \times \mathbf{u}) d\forall$, enstrophy $\int_s |\boldsymbol{\omega}| dS = \int_s |\nabla \times \mathbf{u}| dS$, vortex pairing, isotropic, homogeneous, and stationary turbulence, the vortex sheet, the vortex line, the vortex tube, the vortex ring, the vortex filament, the vortex blub, and different types of vortices like the Gortler vortex, the hairpin and horseshoe vortices, and the Moffatt vortex.

NOTE

1. Actually, vorticity is a pseudovector.

Bibliography

1. Shomali, Z., Ghazanfarian, J., Abbassi, A. Effect of film properties for non-linear DPL model in a nanoscale MOSFET with high-k material: $ZrO_2/HfO_2/La_2O_3$. Superlattices and Microstructures. **83**:699–718 (2015).

2. Bondi, H. Negative mass in general relativity. Reviews of Modern Physics. **29**: 423 (1957).

3. Balaj, M., Roohi, E., Mohammadzadeh, A. Regulation of anti-Fourier heat transfer for non-equilibrium gas flows through micro/nanochannels. International Journal of Thermal Sciences. **118**:24–39 (2017).

4. Schmidt, M., Kusche, R., Hippler, T., Donges, J., Kronmuller, W., von Issendorff, B., Haberland, H., Negative heat capacity for a cluster of 147 sodium atoms. Physical Review Letters. **86**(7):1191 (2001).

5. Shliomis, M.I., Morozov, K.I. Negative viscosity of ferrofluid under alternating magnetic field. Physics of Fluids. **6**:2855–2861 (1994).

6. Lebrun, P. Superfluid helium cryogenics for the large Hadron collider project at CERN. Cryogenics. **34**:1–8 (1994).

7. Volovik, G.E. Superfluid analogies of cosmological phenomena. Physics Reports. **351**(4):195–348 (2001).

8. Lpez, H.M., Gachelin, J., Douarche, C., Auradou, H., Clment, E. Turning bacteria suspensions into superfluids. Physical Review Letters. **115**:028301 (2015).

9. Todd, B.D., Evans, D.J., Daivis, P.J. Pressure tensor for inhomogeneous fluids. Physical Review E. **52**(2):1627 (1995).

10. Kozachok, A. Justification of Stokes hypothesis and Navier–Stokes first exact transformation for viscous compressible fluid. European Journal of Mechanics-B/Fluids. **72**:701–705 (2018).

11. Buresti, G. A note on Stokes hypothesis. Acta Mechanica. **226**:3555–3559 (2015).

12. Gad-el Hak, M., Bandyopadhyay, P.R. Questions in fluid mechanics. Journal of Fluids Engineering. **117**(3):5 (1995).

13. White, F.M. Viscous Fluid Flow, 3rd edition, McGraw-Hill (2006).

14. Liebermann, L.N. The second viscosity of liquids. Physical Review. **75**(9):1415 (1949).

15. Cousins, D.J., Enrico, M.P., Fisher, S.N., Pickett, G.R., Shaw, N.S. Dynamics with a non-Newtonian gas: The force on a body moving through a beam of excitations in superfluid 3 He. Physical Review Letters. **79**(12):2285 (1997).

16. Ouazzi, A., Turek, S. Numerical methods and simulation techniques for flow with shear and pressure dependent viscosity. In Numerical mathematics and advanced applications (pp. 668–676). Springer, Berlin, Heidelberg (2004).

17. Ouazzi, A., Turek, S., Hron, J. Finite element methods for the simulation of incompressible powder flow. Communications in Numerical Methods in Engineering. 21(10):581–596 (2005).

18. Koeller, R.C. Applications of fractional calculus to the theory of viscoelasticity. Journal of Applied Mechanics. 51(2):299–307 (1984).

19. Di Paola, M., Failla, G., Pirrotta, A. Stationary and non-stationary stochastic response of linear fractional viscoelastic systems. Probabilistic Engineering Mechanics. 28:85–90 (2012).

20. Hashin Z. Viscoelastic behavior of heterogeneous media. Journal of Applied Mechanics. 32(3):630–636 (1965).

21. Chen, X. Numerical modeling of fluid-structure interaction with rheologically complex fluids. Doctoral dissertation, Technische Universitat (2014).

22. Korteweg, D.J. Sur la forme que prennent les equations du mouvement des fluides Si ion tient compte des forces capillaires causees par des variations de densite considerables mais continues et sur la theorie de la capillarite dans lhypoth'ese dune variation continue de la densite. Archives Neerlandaises des Sciences Exactes et Naturelles. 6:1–24 (1901).

23. Zhou, Y., Peng, L. On the time-fractional Navier–Stokes equations. Computers & Mathematics with Applications. 73(6):874–891 (2017).

24. Eringen, A.C. On nonlocal fluid mechanics. International Journal of Engineering Science. 10(6):561–575 (1972).

25. Van, P., Fulop, T.: Weakly non-local fluid mechanics: The Schrodinger equation. Proceedings of the Royal Society A: Mathematical, Physical and Engineering Sciences. 462:541–557 (2006).

26. Nobari, M.R.H., Ghazanfarian, J. A numerical investigation of fluid flow over a rotating cylinder with cross flow oscillation. Computers & Fluids. 38:2026–2036 (2009).

27. Ghazanfarian, J., Nobari, M.R.H. A numerical study of convective heat transfer from a rotating cylinder with cross-flow oscillation. International Journal of Heat and Mass Transfer. 52:5402–5411 (2009).

28. Nobari, M.R.H., Ghazanfarian, J. Convective heat transfer from a rotating cylinder with inline oscillation. International Journal of Thermal Science. 49:2026–2036 (2010).

29. Amiraslanpour, M., Ghazanfarian, J., Razavi, S.E. Drag suppression for 2D oscillating cylinder with various arrangement of splitters at Re=100: A high-amplitude study with OpenFOAM. Journal of Wind Engineering and Industrial Aerodynamics. 164:128–137 (2017).

30. Ghazanfarian, J., Taghilou, B. Active heat transfer augmentation of bundle of tubes by partial oscillatory excitation. Journal of Thermophysics and Heat Transfer. **32**:590–604 (2018).

31. Ghazanfarian, J. Introducing the sliding-wall concept for heat transfer augmentation: The case of flow over square cylinder at incidence. Heat Transfer Engineering. **41**:751–764 (2020).

32. Ghazanfarian, J., Ghanbari, D.: Computational fluid dynammics investigation of turbulent flow inside a rotary double external gear pump, Journal of Fluids Engineering-Transactions of ASME, **137**(2), 021101 (2015).

33. Haleh, S., Shomali, Z., Ghazanfarian J. Combined active-passive heat transfer control using slotted fins and oscillation: The cases of single cylinder and tube bank. International Journal of Heat and Mass Transfer. 182:121972 (2022).

34. Xu, H., Jiang, X., Yu, B. Numerical analysis of the space fractional Navier–Stokes equations. Applied Mathematics Letters. **69**:94–100 (2017).

35. Chemin, J.Y., Desjardins, B., Gallagher, I., Grenier, E. Fluids with anisotropic viscosity. ESAIM: Mathematical Modelling and Numerical Analysis. **34**(2):315–335 (2000).

36. Ghazanfarian, J., Abbassi, A. Mixed convection in a square cavity filled with a porous medium and different exit port position. Journal of Porous Media. **10**:701–718 (2007).

37. Mojtaba, A., Ghazanfarian, J., Nabaei, H., Taleghani, Md. H. Evaluation of laminar airflow heating, ventilation, and air conditioning system for particle dispersion control in operating room including staffs: A non-Boussinesq Lagrangian study. Journal of Building Physics. **45**(2):236–264 (2021).

38. Brackbill, J.U., Kothe, D.B., Zemach, C. A continuum method for modeling surface tension. Journal of Computational Physics. **100**:335–354 (1992).

39. Sachs, W., Meyn, V. Pressure and temperature dependence of the surface tension in the system natural gas/water principles of investigation and the first precise experimental data for pure methane/water at 25 C up to 46.8 MPa. Colloids and Surfaces A: Physicochemical and Engineering Aspects. **94**(2–3):291–301 (1995).

40. Ali, T.A., Khelladi, S., Ramirez, L., Nogueira, X. Cavitation modeling using compressible Navier–Stokes and Korteweg equations. Computational Methods in Multiphase Flow VIII. **89**:425 (2015).

41. Steinhoff, J., Underhill, D. Modification of the Euler equations for "vorticity confinement": Application to the computation of interacting vortex rings. Physics of Fluids. **16**:2738–2744 (1994).

42. Brenner, H. Kinematics of volume transport. Physica A. **349**:11–59 (2005).

43. Brenner, H., Bielenberg, J.R. A continuum approach to phoretic motions: Thermophoresis. Physica A: Statistical Mechanics and its Applications. **355**:251–273 (2005).

44. Finlayson, B.A. Existence of variational principles for the Navier–Stokes equation. Physics of Fluids. **15**(6):963–967 (1972).

45. Yasue, K. A variational principle for the Navier–Stokes equation. Journal of Functional Analysis. **51**(2):133–141 (1983).

46. Bao, W., Jin, S. Weakly compressible high-order I-stable central difference schemes for incompressible viscous flows. Computer Methods in Applied Mechanics and Engineering. **190**:5009–5026 (2001).

47. Lighthill, M.J. On sound generated aerodynamically II. Turbulence as a Source of Sound Proceedings of the Royal Society A: Mathematical, Physical and Engineering Sciences. **222**:1–32 (1954).

48. Curle, N. The Influence of solid boundaries upon aerodynamic sound. Proceedings of the Royal Society A: Mathematical, Physical and Engineering Sciences. **231**:505–510 (1955).

49. Williams, J.E.F., Hawkings, D.L. Sound generation by turbulence and surfaces in arbitrary motion. Philosophical Transactions of the Royal Society A: Mathematical, Physical and Engineering Sciences. **264**:321 (1969).

50. Howe, M.S. Theory of vortex sound (Vol. 33). Cambridge University Press, 2003.

51. Clarke, J.F. Small amplitude gas dynamic disturbances in an exploding atmosphere. Journal of Fluid Mechanics. **89**:343–355 (1978).

52. Conejero, J.A., Lizama, C., Murillo-Arcila, M. On the existence of chaos for the viscous van Wijngaarden–Eringen equation. Chaos, Solitons & Fractals. **89**:100–104 (2016).

53. Sevik, M.M. Topics in hydro-acoustics. In Aero-and Hydro-Acoustics (pp. 285–308). Springer, Berlin, Heidelberg (1986).

54. Doak, P.E. Fundamentals of aerodynamic sound theory and flow duct acoustics. Journal of Sound and Vibration. **28**(3):527–561 (1973).

55. Fahy, F., Gardonio, P. Sound and structural vibration. Radiation, transmission, 2nd edition. Elsevier (2007).

56. Chaudhry, M.H. Applied Hydraulic Transients, 3rd edition. Springer, 2014.

57. Łukaszewicz, G. Micropolar fluids theory and applications, Springer, New York (1999).

58. Szeri, A.Z. Fluid film lubrication: Theory and design. Cambridge University Press, 1998.

59. Rajagopal, K.R., Szeri, A.Z. On an inconsistency in the derivation of the equations of elastohydrodynamic lubrication. Proceedings of the Royal Society of London. Series A: Mathematical, Physical and Engineering Sciences. **459**(2039):2771–2786 (2003).

60. Vinogradova, O.I. Drainage of a thin liquid film confined between hydrophobic surfaces. Langmuir. **11**:2213–2220 (1995).

61. Mangler, W. Boundary layers on bodies of revolution in symmetrical flow. Aerodynamische Versuchsanstalt, Goettingen E.V (1945).

62. Neustupa, J., Penel, P. The Navier–Stokes equation with slip boundary conditions (Mathematical Analysis in Fluid and Gas Dynamics), 46–57 (2007).

63. Niavarani, A., Priezjev, N.V. Modeling the combined effect of surface roughness and shear rate on slip flow of simple fluids. Physical Review E. **81**:011606 (2010).

64. Zeinali, B., Ghazanfarian, J. Turbulent flow over partially superhydrophobic underwater structures: The case of flow over sphere and step. Ocean Engineering. **195**:106688 (2020).

65. Zeinali, B., Ghazanfarian, J., Lessani, B. Janus surface concept for three-dimensional turbulent flows. Computers & Fluids, **170**:213–221 (2018).

66. Ali Rezaei, B., Rezaei Barandagh, A., Ghazanfarian, J. Hybrid heat transfer augmentation from oscillating cylinder with partial superhydrophobic Janus surfaces. Heat Transfer Engineering. **44**:1878–1902 (2022).

67. Kucera, P., Neustupa, J., Penel, P. Navier–Stokes' equation with the generalized impermeability boundary conditions and initial data in domains of powers of the Stokes operator (Kyoto Conference on the Navier–Stokes Equations and their Applications) **1**:237–250 (2007).

68. Bellout, H., Neustupa, J., Penel, P. On the Navier–Stokes equation with boundary conditions based on vorticity. Mathematische Nachrichten. **269**:59–72 (2004).

69. Aris, A. Vectors, tensors, and the basic equations of fluid mechanics. Dover Publications, New York (1989).

70. Saghatchi, R., Ghazanfarian, J. A novel SPH method for the solution of dual-phase-lag model with temperature-jump boundary condition in nanoscale. Applied Mathematical Modelling. **39**(3):1063–1073 (2015).

71. Ghazanfarian, J., Moradi, M. Hybrid SPH-MD two-phase modelling of 3D free-surface flows introducing double K-H instability. Engineering Analysis with Boundary Elements. **88**:115–131 (2018).

72. Saghatchi, R., Ghazanfarian, J., Gorji-Bandpy, M. Numerical simulation of water-entry and sedimentation of an elliptic cylinder using smoothed-particle hydrodynamics method. Journal of Offshore Mechanics and Arctic Engineering-Transactions of ASME. **136**(3):031801 (2014).

73. Ghazanfarian, J., Saghatchi, R., Gorji-Bandpy, M. Turbulent fluid-structure interaction of water-entry/exit of a rotating circular cylinder using SPH method. International Journal of Modern Physics C. **26**(8):1550088 (2015).

74. Ghazanfarian, J., Saghatchi, R., Gorji-Bandpy, M. SPH simulation of turbulent flow past a high-frequency in-line oscillating cylinder near free-surface. International Journal of Modern Physics C. **27**(12):1650152 (2016).

75. Ansari, A., Khavasi, E., Ghazanfarian, J. Experimental and SPH studies of reciprocal wet-bed dam break flow over obstacles, International Journal of Modern Physics C, **32**:2150098 (2021).

76. Ghazanfarian, J., Abbassi, A. Investigation of 2-D transient heat transfer under the effect of dual-phase-lag model in a nanoscale geometry. International Journal of Thermophysics. **33**:552–566 (2012).

77. Myint-U, T., Debnath, L. Linear partial differential equations for scientists and engineers. Springer (2007).

78. Arrigo, D.J. Analytical techniques for solving nonlinear partial differential equations. Synthesis Lectures on Mathematics and Statistics. **11**(3):1–165 (2019).

79. Panton, T.L. Incompressible flow. 4th edition, Wiley, Hoboken, New Jersey (2013).

80. Schlichting. H. Berechnung ebener periodischer Grenzschichtstrornungen, Phvs. Z., **33**:327–335 (1932).

81. Ghazanfarian, J., Shomali, Z., Abbassi, A. Macro-to-nanoscale heat and mass transfer: The lagging behavior. International Journal of Thermophysics. **36**: 1416–1467 (2015).

82. Shomali, Z., Kovacs, R., Van, P., Kudinov, I. V., Ghazanfarian, J. Recent progresses and future directions of lagging heat models in thermodynamics and bioheat transfer. Continuum Mechanics and Thermodynamics. **34**:637–679 (2022).

83. Shomali, Z., Abbassi, A., Ghazanfarian, J. Development of non-Fourier thermal attitude for three-dimensional and graphene-based MOS devices, Applied Thermal Engineering. **104**:616–627 (2016).

84. Minkowycz, W.J., Haji-Sheikh, A., Vafai, K.F. On departure from local thermal equilibrium in porous media due to a rapidly changing heat source: The sparrow number. International Journal of Heat and Mass Transfer. **42**(18):3373–3385 (1999).

85. Basirat, H., Ghazanfarian, J., Forooghi., P. Implementation of dualphase-lag model at different Knudsen numbers within slab heat transfer. In Proceedings of the International Conference on Modeling and Simulation (MS06), Konia, Turkey, Vol. 895. (2006).

86. Wang, H., Guo, Z. Thermon gas as the thermal energy carrier in gas and metals. Chinese Science Bulletin. **55**:3350–3355 (2010).

87. Yang, R., Chen, G., Laroche, M., Taur, Y. Simulation of nanoscale multidimensional transient heat conduction problems using ballistic-diffusive equations and phonon Boltzmann equation. Journal of Heat Transfer. **127**:298–306 (2005).

88. Guyer, R.A., Krumhansl, J.A. Solution of the linearized phonon Boltzmann equation. Physical Review. **148**:766 (1966).

89. Vermeersch B., Carrete J., Mingo N., Shakouri A. Superdiffusive heat conduction in semiconductor alloys. I. Theoretical foundations. Physical Review B. **91**(8):085202 (2015).

90. Christov, C.I. On frame indifferent formulation of the Maxwell–Cattaneo model of finite-speed heat conduction. Mechanics Research Communications. **36**:481–486 (2009).

91. Khayat, R.E., deBruyn, J., Niknami, M., Stranges, D.F., Khorasany, R.M.H. Non-Fourier effects in macro- and micro-scale non-isothermal flow of liquids and gases. International Journal of Thermal Sciences. **97**:163–177 (2015).

92. Chung, T.J. Applied Continuum Mechanics, Cambridge University Press, New York (1996).

93. Liepmann, H.W., Roshko, A. Elements of gasdynamics. Courier Corporation (2001).

94. Lior, N., Sarmiento-Darkin, W., Al-Sharqawi, H.S. The exergy fields in transport processes: Their calculation and use. Energy. **31**:553–578 (2006).

95. Jou, D., Casas-Vazquez, J., Lebon, G. Extended Irreversible Thermodynamics, 4th edition, Springer, New York (2010).

96. Ghazanfarian, J., Shomali, Z. Investigation of dual-phase-lag heat conduction model in a nanoscale metal-oxide-semiconductor field-effect transistor. International Journal of Heat and Mass Transfer. **55**(21–22): 6231–6237 (2012).

97. Bejan, A. Convective heat transfer, 4th edition, Wiley, Hoboken, New Jersey (2013).

98. Bouzari, S., Ghazanfarian., J. Unsteady forced convection over cylinder with radial fins in cross flow. Applied Thermal Engineering. 112:214–225 (2017).

99. Jackson, J.D. Classical Electrodynamics, 3rd edition, Wiley (1999).

100. Arbab, A.I. The analogy between electromagnetism and hydrodynamics. Physics Essays. **24**(2):254 (2011).

101. Moghaddam, M., Ghazanfarian, J., Abbassi, A. Implementation of DPL-DD model for the simulation of nanoscale MOS devices. IEEE Transactions on Electron Devices. **61**(9):3131–3138 (2014).

102. Masood, M., Abbassi, A., Ghazanfarian., J, Microstructure effects on performance and deactivation of hierarchically structured porous catalyst: A pore network model. arXiv preprint arXiv:2202.02720 (2022).

103. Wachutka, G.K. Rigorous thermodynamic treatment of heat generation and conduction in semiconductor device modeling. IEEE Transactions on Computer-Aided Design of Integrated Circuits and Systems. **9**:1141–1149 (1990).

104. Ghazanfarian, J., Mohammad Mostafa M., Kenji U. Piezoelectric energy harvesting: A systematic review of reviews. In Actuators, Vol. 10(12), p. 312. MDPI (2021).

105. Fang, D., Liu, J. Basic equations of piezoelectric materials. Fracture Mechanics of Piezoelectric and Ferroelectric Solids. **4**:77–95 (2013).

106. Landau L.D., Lifshitz E.M. Elecrodynamics of continuous media, Pergamon Press, Oxford, UK (1960).

107. Torkashvand, K., Poursaeidi, E., Ghazanfarian, J. Experimental and numerical study of thermal conductivity of plasma-sprayed thermal barrier coatings with random distributions of pores, Applied Thermal Engineering. **137**:494–503 (2018).

108. Ergun, S., Fluid flow through packed columns. Chemical Engineering Progress. **48**:89–94 (1952).

109. Klinkenberg, L.J. The permeability of porous media to liquids and gases. Drilling and Production Practice. American Petroleum Institute, 200–213 (1941).

110. Brinkman, H.C. A calculation of the viscous force exerted by a flowing fluid on a dense swarm of particles. Applied Scientific Research. **1**:27–34 (1949).

111. Ovaysi, S., Piri, M. Direct pore-level modeling of incompressible fluid flow in porous media. Journal of Computational Physics. 229(19):7456–7476 (2010).

112. Nabovati, A., Sousa, A.C.M. Fluid flow simulation in random porous media at pore level using lattice Boltzmann method. In New Trends in Fluid Mechanics Research, Springer, Berlin, Heidelberg, pp. 518–521 (2007).

113. Moghaddam, M., Abbassi, Abbas, Ghazanfarian, J., Jalilian, S. Investigation of microstructure effects on performance of hierarchically structured porous catalyst using a novel pore network model. Chemical Engineering Journal. 388:124261 (2020).

114. Lemaitre, R., Adler, P.M. Fractal porous media IV: Three-dimensional Stokes flow through random media and regular fractals. Transport in Porous Media. 5(4), 325–340 (1990).

115. Vafai, K., Tien, C.L. Boundary and inertia effects on flow and heat transfer in porous media. International Journal of Heat and Mass Transfer. 24(2):195–203 (1981).

116. Liu, Q., Vasilyev, O.V. A Brinkman penalization method for compressible flows in complex geometries. Journal of Computational Physics. 227(2):946–966 (2007).

117. Lighthill, M.J., Whitham, G.B. On kinematic waves II. A theory of traffic flow in long crowded roads. Proceedings of the Royal Society A. **229**:317–345 (1955).

118. Richards, P.I. Shockwaves on the highway. Operations Research, 4:42–51 (1956).

119. Aw, A.A., Rascle, M. Resurrection of "second order" models of traffic flow. SIAM Journal on Applied Mathematics. **60**:916–938 (2000).

120. Vermolen, F.J., Gharasoo, M.G., Zitha, P.L., Bruining, J. Numerical solutions of some diffuse interface problems: The Cahn–Hilliard equation and the model of Thomas and Windle. International Journal for Multiscale Computational Engineering. 7(6) (2009).

121. Allen, S.M. Cahn, J.W. Ground state structures in ordered binary alloys with second neighbor interactions. Acta Metallic. **20**(3):423–433 (1972).

122. Vallis, G.K. Atmospheric and oceanic fluid dynamics. Cambridge University Press, 2017.

123. De Bortoli, A.L., Andreis, G.S.L., Pereira, F.N. Modeling and simulation of reactive flows. Elsevier (2015).

124. Peters, N.: Turbulent combustion, Cambridge University Press, (2001).

125. Jager, W., Rannacher, R., Warnatz, J. Reactive flows, diffusion and transport: From experiments via mathematical modeling to numerical simulation and optimization. Berlin/Heidelberg, Springer (2006).

126. Peters, N.: Turbulent combustion, Cambridge University (1941).

127. Maxey, M.R., Patel, B.K. Localized force representations for particles sedimenting in Stokes flow. International Journal of Multiphase Flow. **27**:1603–1626 (2001).

128. Cowin, S.C. The theory of polar fluids. Advances in Applied Mechanics. **14**:279–347 (1974).

129. Henriques, F.C. Studies of thermal injury. V. The prechoosing the parameter values for governing equations dictability and the significance of thermally induced rate processes leading to irreversible epidermal injury. Archives of Pathology. **43**:489–502 (1947).

130. Pennes, H.H. Analysis of tissue and arterial blood temperature in the resting human forearm. Journal of Applied Physics. **1**:93–102 (1948).

131. Ghazanfarian, J., Saghatchi, R., Patil, D.V. Implementation of smoothed-particle hydrodynamics for non-linear Pennes' bioheat transfer equation. Applied Mathematics and Computation. **259**:21–31 (2015).

132. Chen, M.M., Holmes, K.R. Microvascular contributions in tissue heat transfer. Annals of the New York Academy of Sciences. **335**:137–150 (1980).

133. Weinbaum, S., Jiji, L.M., Lemons, D.E. Theory and experiment for the effect of vascular microstructure on surface tissue heat transfer. Part I. Anatomical foundation and model conceptualization. ASME Journal of Biomechanical Engineering. **106**:321–330 (1984).

134. Weinbaum, S., Jiji, L.M. A new simplified bioheat equation for the effect of blood flow on local average tissue temperature. ASME Journal of Biomechanical Engineering. **107**:131–137 (1985).

135. Zhou, J., Chen, J.K., Zhang, Y. Dual-phase lag effects on thermal damage to biological tissues caused by laser irradiation. Computers in Biology and Medicine. **39**:286–293 (2009).

136. Jamshidi, M., Ghazanfarian, J. Blood flow effects in thermal treatment of three-dimensional non-Fourier multi-layered skin structure. Heat Transfer Engineering. **42**:929–946 (2021).

137. Jamshidi, M., Ghazanfarian, J. Dual-phase-lag analysis of $CNTMoS_2$- ZrO_2-SiO_2-Si nano-transistor and arteriole in multi-layered skin. Applied Mathematical Modelling. **60**:490–507 (2018).

138. Katopodes, N.D. Free-surface flow. Volume of Fluid Method. Elsevier, pp. 766–802 (2019).

139. Williams F.A. Turbulent combustion. In The Mathematics of Combustion, Society for Industrial and Applied Mathematics. pp. 197–1318 (1985).

140. Plesset, M.S. The dynamics of cavitation bubbles. Journal of Applied Mechanics. **16**:228–231 (1949).

141. Buckley, S.E., Leverett, M.C. Mechanism of fluid displacements in sands. Transactions of the AIME. **146**:107–116 (1942).

142. Carey, V.P. Liquid vapor phase change phenomena: An introduction to the thermophysics of vaporization and condensation processes in heat transfer equipment. 2nd Edition, Taylor & Francis, (1984).

143. Ghazanfarian, J., Abbassi, A. Heat transfer and fluid flow in microchannels and nanochannels at high Knudsen number using thermal lattice-Boltzmann method. Physical Review E. **82**:026307 (2010).

144. Succi, S. The lattice Boltzmann equation: For fluid dynamics and beyond. Oxford University Press (2001).

145. Dolbeault, J. Kinetic models and quantum effects: A modified Boltzmann equation for Fermi–Dirac particles. Archive for Rational Mechanics and Analysis. **127**(2):101–131 (1994).

146. Ghazanfarian, J., Abbassi, A. Effect of boundary phonon scattering on dual-phase lag model to simulate micro- and nano-scale heat conduction. International Journal of Heat and Mass Transfer. **52**:3706–3711 (2009).

147. Karniadakis, G., Beskok, A., Aluru, N. Microflows and nanoflows fundamentals and simulation. Springer, New York (2005).

148. Shomali, Z., Pedar, B., Ghazanfarian, J., Abbassi, A., Monte-Carlo. Parallel simulation of phonon transport for 3D silicon nano-devices. International Journal of Thermal Sciences. **114**:139–154 (2017).

149. Samian, R.S., Abbassi, A., Ghazanfarian, J. Thermal investigation of common 2D FETs and new generation of 3D FETs using Boltzmann transport equation in nanoscale. International Journal of Modern Physics C. **24**(9):1350064 (2013).

150. Samian, R.S., Abbassi, A., Ghazanfarian, J., Transient conduction simulation of a nanoscale hotspot using finite volume lattice Boltzmann method. International Journal of Modern Physics C. **25**(4):1350103 (2014).

151. Wyatt, R.E. Quantum dynamics with trajectories: Introduction to quantum hydrodynamics. Springer (2006).

152. Deb, B.M., Ghosh, S.K. Schrodinger fluid dynamics of many-electron systems in a time-dependent density-functional framework. The Journal of Chemical Physics. **77**(1):342–348 (1982).

153. Kobayashi, M., Parnaudeau, P., Luddens, F., Lothod, C., Danaila, L., Brachet, M., Danaila, I. Quantum turbulence simulations using the Gross–Pitaevskii equation: High-performance computing and new numerical benchmarks. Computer Physics Communication. **258**:107579 (2020).

154. Davies, C.N. Definitive equations for the fluid resistance of spheres. Proceedings of the Physical Society. **57**:259 (1945).

155. Alexander, F., Garcia, A.L., Alder, B.J. Direct simulation Monte Carlo for thin-film bearings. Physics of Fluids. **6**(12):3854–3860 (1994).

156. Ghazanfarian, J., Shomali, Z., Xiong, Sh. Nanoscale energy transport, 21st Century Nanoscience, A Handbook. Vol. 1, Taylor & Francis Press (2019).

157. Dumitru, B., Kai, D., Enrico, S.,Trujillo, J.J. Fractional calculus: Models and numerical methods. World Scientific, (2012).

158. Wheatcraft, S.W., Meerschaert, M.M. Fractional conservation of mass. Advances in Water Resources. **31**:1377–1381 (2008).

159. Momani, S., Odibat, Z. Analytical solution of a time-fractional Navier-Stokes equation by Adomian decomposition method. Applied Mathematics and Computation. **177**:488–494 (2006).

160. Metzler, R., Klafter, J. The random walk's guide to anomalous diffusion: A fractional dynamics approach. Physics Reports. **339**:1–77 (2000).

161. Holm, S., Nsholm, S.P. Comparison of fractional wave equations for power law attenuation in ultrasound and elastography. Ultrasound in Medicine & Biology. **40**:695–703 (2014).

162. Laskin, N. Fractional Schrodinger equation. Physical Review E. **66**(5):056108 (2002).

163. Gouesbet, G., Weill, M.E. Complexities and entropies of periodic series with application to the transition to turbulence in the logistic map. Physical Review A. **30**(3):1442 (1984).

164. Kardar, M., Parisi, G., Zhang, Y.-C. Dynamic scaling of growing interfaces. Physical Review Letters. **56**(9):889–892 (1986).

165. Gupta, M., Mukhopadhyay, S. A study on generalized thermoelasticity theory based on non-local heat conduction model with dual-phase-lag. Journal of Thermal Stresses. **42**:1123–1135 (2019).

166. Atzberger, P.J. Stochastic Eulerian Lagrangian methods for fluid structure interactions with thermal fluctuations. Journal of Computational Physics. **230**(8):2821–2837 (2011).

167. Da Prato, G., Debussche, A., Temam, R. Stochastic Burgers' equation. Nonlinear Differential Equations and Applications NoDEA. **1**(4):389–402 (1994).

168. Hofmanova, M., Breit, D. Stochastic Navier–Stokes equations for compressible fluids. Indiana University Mathematics Journal. **65**(4):1183–1250 (2016).

169. Da Prato, G., Debussche, A. Stochastic Cahn–Hilliard equation. Nonlinear Analysis: Theory, Methods & Applications. **26**(2):241–263 (1996).

170. Li, A., Ahmadi, G. Dispersion and deposition of spherical particles from point sources in a turbulent channel flow. Aerosol Science and Technology. **16**:209–226 (1992).

171. De la Pena, L., Cetto, A.M. The quantum dice: An introduction to stochastic electrodynamics, Springer (2013).

172. Ali, Y.M., Zhang, L.C. Relativistic heat conduction. International Journal of Heat Mass Transfer. **48**:2397 (2005).

173. Weinberg, S. Gravitation and Cosmology: Principles and Applications of the General Theory of Relativity, Wiley (2013).

174. Jackiw, R., Nair, V.P., Pi, S.-Y., Polychronakos, A.P. Perfect fluid theory and its extensions. Journal of Physics A: Mathematical and General. **37**:R327–R432 (2004).

175. Kambe, T. New scenario of turbulence theory and wall-bounded turbulence: Theoretical significance. Geophysical & Astrophysical Fluid Dynamics. **111**:448–507 (2017).

176. Ledvina, S.A., Ma, Y.-J., Kallio, E. Modeling and simulating flowing plasmas and related phenomena. Space Science Reviews. **139**(1–4):143 (2008).

177. Marmanis, H. Analogy between the Navier–Stokes equations and Maxwell's equations: Application to turbulence. Physics of Fluids. **10**(6):1428–1437 (1998).

178. Baskurt, O.K., Meiselman, H.J. Blood rheology and hemodynamics. Seminars in thrombosis and hemostasis. **29**(05):435–450 (2003).

179. Alexandre, A. Dark fluid: A complex scalar field to unify dark energy and dark matter. Physical Review D. **74**(4):043516 (2006).

180. Gautier, N., Aider, J.L., Duriez, T., Noack, B.R., Segond, M., Abel, M. Closed-loop separation control using machine learning. Journal of Fluid Mechanics. **770**:442–457 (2015).

181. Duriez, T., Brunton, S.L., Noack, B.R. Machine learning control-taming nonlinear dynamics and turbulence. Springer (2017).

182. Tsang, A.C.H., Tong, P.W., Nallan, S., Pak, O.S. Self-learning how to swim at low Reynolds number. Physical Review Fluids. **5**:074101 (2020).

183. Kamyar, D., Ghazanfarian. J. Active control of flow over rotating cylinder by multiple jets using deep reinforcement learning. arXiv preprint arXiv:2307.12083 (2023).

184. Croicu, A.M., Hussaini, M.Y. Multiobjective stochastic control in fluid dynamics via game theory approach: Application to the periodic Burgers equation. Journal of Optimization Theory and Applications. **139**(3):501–514 (2008).

185. Periaux, J., Chen, H.Q., Mantel, B., Sefrioui, M., Sui, H.T. Combining game theory and genetic algorithms with application to DDM-nozzle optimization problems. Finite Elements in Analysis and Design. **37**(5):417–429 (2001).

186. Iida, H., Nakagawa, T., Spoerer, K. Game information dynamic models based on fluid mechanics. Entertainment Computing. **3**(3):89–99 (2012).

187. Ghazanfarian, J., Khavasi, E., Yousefi, H., Amiraslanpour, M., Teymouri, S., Bayat, R. Modern online learning tools over the platform of virtual/augmented reality. In Optimizing Student Engagement in Online Learning Environments (pp. 101–126). IGI Global (2018).

188. Goldburg, W.I., Cerbus, R.T. Turbulence as Information. Physical Review E. **88**:053012 (2013).

189. Zhang, P., Rosen, M., Peterson, S.D., Porfiri, M. An information theoretic approach to study fluid-structure interactions. Journal of Fluid Mechanics. **848**:968–986 (2018).

190. Baudrimont, R. Entropic description of gravity by thermodynamics relativistic fluids and the information theory (2019).

191. Foster, N., Fedkiw, R. Practical animation of liquids. In Proceedings of the 28th annual conference on computer graphics and interactive techniques (2001).

192. Branlard, E. A brief introduction to vortex methods, wind turbine aerodynamics and vorticity-based methods. 483–492 (2017).

193. Thomson, W. On vortex motion. Earth and Environmental Science Transactions of the Royal Society of Edinburgh. **25**:217–260 (1868).

194. Garth, C., Tricoche, X., Salzbrunn, T., Bobach, T., Scheuermann, G. Surface techniques for vortex visualization. VisSym. **4**:155–164 (2004).

195. Holmen, V. Methods for vortex identification. Theses in Mathematical Sciences (2012).

196. Gao, Y., Liu, C. Rortex and comparison with eigenvalue-based vortex identification criteria. Physics of Fluids. 30(8):085107 (2018).

197. McCormack, P. Vortex, Molecular spin and nanovorticity: An introduction. Springer (2011).

198. Joseph, D.D. Viscous potential flow. Journal of Fluid Mechanics. **479**:191–197 (2003).

Index

Printed in the United States
by Baker & Taylor Publisher Services